International Interdisciplinary Conference on Human Factors, Design and Education for Sustainability 2023

IOP Conference Series: Earth and Environmental Science
Volume 1256

Bandung, Indonesia
14-15 August 2023

Editors:

Carles Sitompul
Clara Theresia

ISBN: 978-1-7138-8193-3
ISSN: 1755-1307

Printed from e-media with permission by:

Curran Associates, Inc.
57 Morehouse Lane
Red Hook, NY 12571

Some format issues inherent in the e-media version may also appear in this print version.

Institute of Physics
Dirac House, Temple Back
Bristol BS1 6BE UK

Phone: 44 1 17 929 7481
Fax: 44 1 17 920 0979

techtracking@iop.org

Additional copies of this publication are available from:

Curran Associates, Inc.
57 Morehouse Lane
Red Hook, NY 12571 USA
Phone: 845-758-0400
Fax: 845-758-2633
Email: curran@proceedings.com
Web: www.proceedings.com

TABLE OF CONTENTS

PREFACE

It is with great honor to present the proceeding of the International Interdisciplinary Conference on Human Factors, Design and Education for Sustainability (ICoHDES) held in Parahyangan Catholic University (UNPAR) Bandung-Indonesia, 14 – 15th of August 2023. The conference is co-supported by the Erasmus+ Capacity Building for Higher Education (CBHE) project called IN2FOOD: Resolving Societal Challenge: Interdisciplinary Approach Towards Fostering Collaborative Innovation in Food Waste Management, Reference Number: 618717-EPP-12020-1-ID-EPPKA2-CBHE-JP.

This proceeding consists 7 reviewed articles which best suited to be published in the scope for IOP Proceeding Series: Earth and Environmental Sciences (EES). Other articles presented in the conference are submitted for a special edition in an Indonesian accredited journal. We are hopeful that the conference and its publications will bring impactful results to the society, bringing together a network of people to resolve sustainability issues through collaborative innovation.

We would like to extend our deepest gratitude to distinguished keynote speakers, namely: Prof. Elina Närvänen (Tampere University, Finland), Prof. Lieven de Marez (Ghent University, Belgium), Dr. (cand)., Anna de Visser-Amundson (Hoteschool The Hague, the Netherlands), Dr. Thedy Yogasara (Parahyangan Catholic University, Indonesia) and Dr. Pius Sugeng Prasetyo (Parahyangan Catholic University, Indonesia). Lastly, we would like to express our gratitude to all participants, our organizing and scientific committee for their relentless supports ensuring the success of this conference.

Best regards,

Carles Sitompul (Conference Chair and Editor)
Clara Theresia (Editor)

Steering Committee:
Johanna Renny Octavia (Parahyangan Catholic University, Indonesia)
Elina Närvänen (Tampere University, Finland)
Nina Mesiranta (Tampere University, Finland)
Jelle Saldien (Ghent University, Belgium)
Anna de Visser-Amundson (Hotelschool The Hague, the Netherlands)
Diah Wihardini (Bina Nusantara University, Indonesia)
Stevanus Wisnu Wijaya (Prasetiya Mulya University, Indonesia)
Eddy Yusuf (Pembangunan Jaya University, Indonesia)
Mustika Tarra (Ma Chung University, Indonesia).

Organizing Committte:
Carles Sitompul (Conference Chair, Parahyangan Catholic University, Indonesia)
Vrameswari Omegawati (Conference Co-Chair, Parahyangan Catholic University, Indonesia)
Clara Theresia (Secretary and Treasurer, Parahyangan Catholic University, Indonesia)
Giovano Alberto (program and Paper Chair, Parahyangan Catholic University, Indonesia)
Icha Wiharini (Social Chair, Parahyangan Catholic University, Indonesia)
Daniel Siswanto (Documentation and Logistics Chair, Parahyangan Catholic University, Indonesia).

Scientific Committee:
Elina Närvänen (Tampere University, Finland)
Nina Mesiranta (Tampere University, Finland)
Hardianto Iristiadi (Bandung Institute of Technology, Indonesia)
Eddy Yusuf (Pembangunan Jaya University, Indonesia)
Markus Hartono (Surabaya University, Indonesia)
Paulina Kus Ariningsih (Parahyangan Catholic University, Indonesia)
Theresia Gunawan (Parahyangan Catholic University, Indonesia)
Vrameswari Omegawati (Parahyangan Catholic University, Indonesia)
Anggia Valerisha (Parahyangan Catholic University, Indonesia)
Yulia Indri Sari (Parahyangan Catholic University, Indonesia)
Sugih Sudharma Tjandra (Parahyangan Catholic University, Indonesia)
Johanna Renny Octavia (Parahyangan Catholic University, Indonesia)
Titi Iswari (Parahyangan Catholic University, Indonesia)
Kinley Aritonang (Parahyangan Catholic University, Indonesia)
Diana Lo (Bina Nusantara University, Indonesia)
Hans Kristianto (Parahyangan Catholic University, Indonesia)
Daniel Siswanto (Parahyangan Catholic University, Indonesia)
Diah Ratna Pratiwi (PAGE Indonesia, UNDP, Indonesia)
Endar Purnawan (Regional Board for National Planning, Indonesia)
Elizabeth A. Dewi (Parahyangan Catholic University, Indonesia)

IOP Conf. Series: Earth and Environmental Science 1256 (2023) 011002 doi:10.1088/1755-1315/1256/1/011002

Peer Review Statement

All papers published in this volume have been reviewed through processes administered by the Editors. Reviews were conducted by expert referees to the professional and scientific standards expected of a proceedings journal published by IOP Publishing.

- Type of peer review: Single anonymous
- Conference submission management system: Morressier
- Number of submissions received: 14
- Number of submissions sent for review: 14
- Number of submissions accepted: 7
- Acceptance Rate (Submissions Accepted/Submission Received x 100): 50%
- Average number of reviews per paper: 1.34
- Total number of reviewers involved: 7

Contact person for queries:

Name: Carles Sitompul

Affiliation: Parahyangan Catholic University

Email: carles@unpar.ac.id

Improving Sustainability Asset Performance Based on Dynamic Data Envelopment Analysis

Muhamad Nugraha and Winda Nur Cahyo*

Master Program in Industrial Engineering, Faculty of Industrial Technology, Islamic University of Indonesia, Jl. Kaliurang, km. 14,5, 55584, Yogyakarta

*)Author's e-mail address: winda.nurcahyo@uii.ac.id

Abstract. The success of an industry is often measured through sustainability performance because it serves as a benchmark of consideration for asset managers in carrying out actions, planning, and decisions. However, changes in the utilization of physical assets between entities impact not only the life cycle of these assets but also sustainability assets that will affect company performance in terms of controlling company efficiency. The DDEA method is an analytical method that can be used to assess the level of efficiency of the performance of sustainability assets. The performance of sustainability assets is measured using variable maintenance costs, asset availability, material consumption as an economic aspect, variable environmental impact costs, employee health costs, and accidents and incidents as social aspects. In addition, material rejects variables, hazardous waste, and energy consumption costs as environmental aspects. From the results of the analysis of sustainability asset performance through the DDEA method in the period 2018–2022, it shows that both before and after organizational changes have no significant impact. This is shown by the level of efficiency of sustainability asset indicators on the impact of organizational change found in DMU 1, DMU 2, DMU 3, DMU 5, DMU 6, DMU 9, and DMU 10. The achievement of constant or efficient conditions with input-oriented CRS efficiency is at a score of 1. While DMU 4, DMU 7, and DMU 8 show increasing conditions with input-oriented CRS efficiency at less than 1.

Keywords: Organizational Change, Asset Sustainability, DDEA, Physical Assets

1. Introduction

In the current era of disruption, not only individual businesses but all industries and business ecosystems are required to be able to compete and survive the conditions of the new business model changes that occur [1], [2]. Various defenses are built by each industry to continue to exist in this era of disruption, including by making changes in the organization to be more agile [3], [4].

This conception is also supported by [5], who argue that more agile organizational changes can increase efficiency and effectiveness. State-owned enterprises (SOEs) of the Republic of Indonesia are also not spared from the impact of disruption, which further changes business structures in various organizations to refocus business [6]. Thus, the industry can overcome previously unforeseen situations to survive in a competitive environment.

The success of an industry is often measured using sustainability performance through social, environmental, and economic indicators [7]–[10]. In addition, sustainability also serves as a benchmark for consideration for asset managers in carrying out actions, planning, and a decision to be proposed [10], [11]. This illustrates that industrial activities that produce a product's economic value also impact society and the safety and health of the surrounding environment. Thus, compliance in asset maintenance, production processes, and manufacturing processes during operations must always be controlled to reduce the impact of industry on the economy, society, and the surrounding environment [8].

Based on the results of an empirical study on one of the manufacturing companies in Indonesia, it is known that in the company's long-term plan for 2020–2024, it has planned to make organizational changes [12]. Based on the asset database until 2022, the company has 1,287 types of physical assets managed by asset managers, and as many as 527 units, or around 41% of the total 1,287 units of existing production machine type assets, are welding machine type assets used for production activities (production assets). However, the utilization of each of the company's assets has also shifted, moving to subsidiaries at around 75%, affiliate companies at around 10%, and those managed by the company's asset managers at only around 15%.

The concept of asset management is widely adopted to optimize decision-making in terms of balancing costs, risks, and performance for asset-intensive companies [13], [14]. However, changes in the utilization of physical assets between different entities certainly have an impact not only on the life cycle but also on the risk of maintaining sustainable assets. This is to ensure safety, availability, and reliability, as well as the provision of products according to the required quality [8].

Based on the impacts of these organizational changes, it is necessary to evaluate how the company performs through the performance of sustainability assets. Researchers have conducted empirical studies on previous studies. [15] in their research through empirical studies using the case exploration method, retested [16] model of 80 sustainability indicators (43 economic aspect indicators, 20 social aspect indicators, and 17 environmental aspect indicators) as a measure of industrial sustainability performance in the manufacturing industry in Italy and Germany. [8]in their research, assessed the impact of maintenance practices on asset sustainability in manufacturing companies located in the German region by measuring the level of efficiency in environmental, social, and economic aspects.

Researchers need to consider the variables conveyed by [8] in evaluating sustainability assets, namely by evaluating other aspects, such as aspects of the environmental budget level and aspects of the health budget level. In addition, it is necessary to consider long-term business conditions. As stated by [17], dynamic conditions greatly affect the company's operational cycle in decision-making. Refocusing business, which has implications for the delegation of the company's main business process and the utilization of production facility assets to subsidiary companies, impacts the maintenance and management activities of sustainable asset organizations so that it can impact company performance for a certain period.

From the consideration of these empirical studies, no results show performance evaluation of sustainability assets from organizational changes in a company by considering dynamic aspects through the Dynamic Data Envelopment Analysis (DDEA) approach. As [17]–[22], dynamic aspects become important to detect efficiency over certain time spans with the DDEA approach. It is considered important because the conditions of organizational change need to be evaluated before and after organizational change as a consideration in decision-making. Thus, researchers consider it necessary to deepen the aspects of sustainability assets that prioritize the evaluation stage of the impact of organizational changes in the company to obtain conclusions that provide benefits, namely by identifying the efficiency of sustainability asset indicators based on DDEA.

2. Research Method

This research was conducted at one of Indonesia's railway industry manufacturing companies. The problem that is the object of research is the increase in sustainability assets in the implementation of maintenance and the impact of organizational changes.

2.1. Data Collection

The data used in this study is sample data in the form of historical data on input and output variables related to improving the performance of sustainability assets in the implementation of maintenance and the impact of organizational changes in the period 2018–2022. The asset transaction data used are the assets of the production facility group for the production process, especially the type of welding machine and sustainability aspects covering economic, social, and environmental aspects.

The determination of the object and period of research is influenced by the reasons for the transaction cycle and methodology, as well as the availability of data that affects the context of the problem. The conditions are described as follows:

- The conditions from 2018 to 2022, the Company achieved the highest fluctuations in company performance, marked by sales achievements, increased asset investment, and organizational changes.
- The phenomenon of organizational disruption, which has recently been introduced massively and was subsequently followed by SOEs of the Republic of Indonesia with significant organizational changes, is shown by the use of assets owned by the Company by Subsidiaries.

2.2. Data Analysis Methods

This research uses quantitative analysis methods through the DDEA approach related to improving the performance of sustainability assets in the implementation of maintenance against the impact of organizational change. The DDEA method in this study aims to evaluate efficiency in welding machine asset maintenance practice variables through a sustainable indicator approach as input and production amount as output variables in short-term and long-term situations. A measure of efficiency [23], [24], described in the maximum ratio of output weights to input weights provided that the ratio for each Decision-Making Unit (DMU) is less than or equal to one (≤ 1). The formula is explained by [20] as follows:

$$\max h_0 = \frac{\sum_{r=1}^{s} u_r y_{ro}}{\sum_{i=1}^{m} v_i x_{io}}; \quad \text{Subject to: } \frac{\sum_{r=1}^{s} u_r y_{rj}}{\sum_{i=1}^{m} v_i x_{ij}} \leq 1 \rightarrow j = 1, \dots, n. \tag{1}$$

$$v_r, u_i \geq 0; \rightarrow r = 1, \dots, s; \text{ and } i = 1, \dots, m \tag{2}$$

Formula (1)(2) above for y_{rj}, x_{ij} is the output and input values to j of the DMU and must be positive. While $u_r, v_i \geq 0$ is the weight of the variable that will determine the solution of the problem. The DDEA model in this study uses a model delivered [20], which is input-oriented to minimize input concerning fixed output. Still, adjustments are made to input variables according to the results of identifying sustainability indicators conveyed [8], [15] so that the model used in this study becomes as shown in Figure 1.

In addition, DDEA Model in this study is input-oriented to minimize input. The dynamic system referred to in this study is related to time, as assumed below:

$$\frac{1}{E_0} = min\left(\sum_{i=1}^{m} v_i X_{io} + \sum_{f=1}^{g} w_f Z_{fo}^{(0)}\right); \text{ Subject to: } \sum_{r=1}^{s} u_r Y_{ro} + \sum_{f=1}^{g} w_f Z_{fo}^{(p)} = 1 \tag{3}$$

$$\left(\sum_{i=1}^{m} v_i X_{ij} + \sum_{f=1}^{g} w_f Z_{fj}^{(0)}\right) - \left(\sum_{r=1}^{s} u_r Y_{rj} + \sum_{f=1}^{g} w_f Z_{fj}^{(0)}\right) \geq 0 \tag{4}$$

$$\left(\sum_{i=1}^{m} v_i X_{ij}^{(t)} + \sum_{f=1}^{g} w_f Z_{fj}^{(t-1)}\right) - \left(\sum_{r=1}^{s} u_r Y_{rj}^{(t)} + \sum_{f=1}^{g} w_f Z_{fj}^{(t)}\right) \geq 0 \tag{5}$$

$$U_r, v_i, w_f \geq \varepsilon, i = 1, \dots, m; \rightarrow r = 1, \dots, s; \ j = 1, \dots, n \tag{6}$$

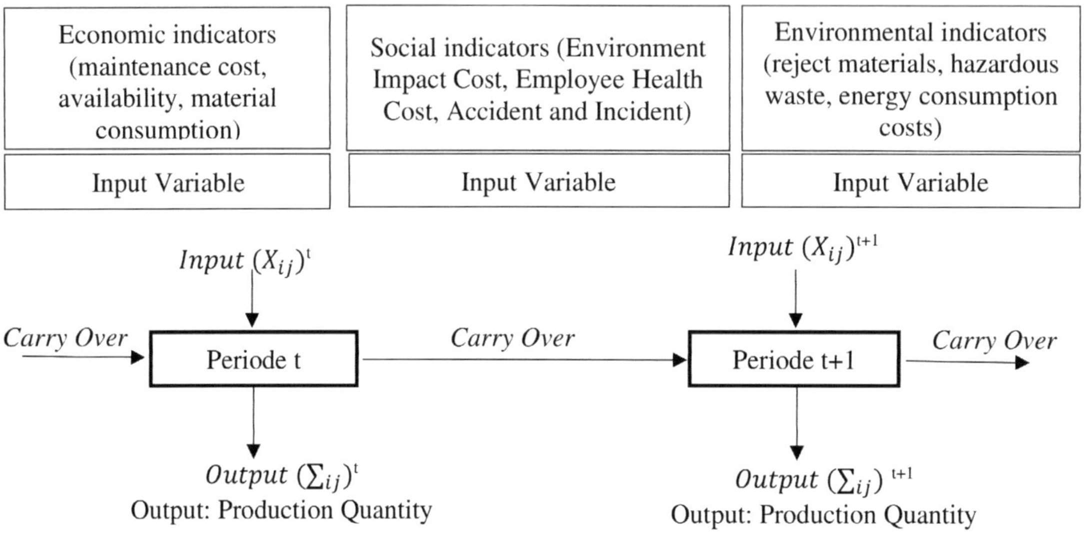

Figure 1. DDEA Model in sustainability aspects

Where u_r, v_i and w_f is multiplication of each output as *r*, input as i, and carry over to f. This model is an input-oriented model that attempts to minimize input concerning a fixed output. The input-output variable for sustainable assets is the sustainability pillar variable in the economic, social, and environmental Triple Bottom Line (TBL) [25]. From the results of identifying the efficiency level of sustainability assets that have been carried out based on the DDEA, an analysis related to slack analysis is carried out. It is to see the potential for slack that occurs in sustainability variables that have been identified. The analysis results identify the level of efficiency and slack analysis, then analyze the input targets on each variable of sustainability assets to see the condition of improving the performance of these sustainability assets. More specifically, observations are made on some sustainability indicators, as in Table 1:

Table 1. Variable units of research input and output

Sustainability Indicator	Variable	Unit
Input (X)		
Economic (IEc)	Costs of Maintenance	Rp.
	Availability	%
	Material Consumption	Pcs/WI
Social (ISo)	Environmental Impact	Rp.
	Costs of Health	Rp.
	Incident & Accident	%
Environment (IEn)	Material Reject	%
	Hazardous Waste	%
	Energy Consumption	Rp
Output (Y)		
Product Delivery	Production amount (Unit)	∑Product

Description of operational variable in this research is follows:

- Costs of Maintenance are variable inputs: the number of accumulated costs of maintenance activities covering preventive and corrective maintenance activities on welding machine assets for production activities in the period 2018–2022, obtained from the company's financial statements contained in the application system in Rupiah (IDR).

- Availability is the percentage (%) input variable of the comparison between available time reduced by downtime divided by the time available on the welding machine in the time range from 2018 to 2022.
- Material Consumption is an input variable for the use of raw materials based on the working instruction report, namely per pcs component/Working Instruction (pcs/WI) in the period 2018 to 2022.
- Environmental Impact is an input variable of the amount of budget value of cost realization for the allocation of activities on environmental impacts in the period 2018–2022 in Rupiah (IDR).
- Costs of health are variable inputs to the amount of budget value realized for the allocation of employee health per employee in the period 2018–2022 in Rupiah (IDR).
- Incident & Accident is an input variable of events caused by activities on company assets from 2018 to 2022.
- Material Reject is an input variable percentage ratio (%) of the number of raw materials rejected or defective in the production process compared to the amount of material to be used in the period 2018–2022.
- Hazardous waste is a variable input level of toxic and hazardous materials waste generated from production process activities in the period 2018–2022 in percentage (%).
- Energy consumption is a variable input level of energy used to carry out production activities and workshop operations in the period 2018–2022 in Rupiah (IDR).
- Production Amount is the variable output of the number of train-type products completed by the company from 2018 to 2022.

Data processing to obtain useful results in research discussions is done using DEA Frontier application software from http://www.deafrontier.net/.

2.3. Assumptions and Limitations of Research

The focus of this research covers several things that are limited to the problem, namely This research uses secondary and primary data on production facility assets in the company, especially welding machine-type assets, in the period 2018–2022. While the scope of assets used as the object of research is welding machine assets whose utilization is carried out by subsidiary companies, which has an impact on efficiency.

The sustainability indicators used in this study depend on the availability of data, so removing some sustainability indicators that have been studied in previous studies tailored to the needs of the company, While the data used in the research is data on the SAP application system, which has the information needed to perform analysis, This makes it possible to state that the resulting conclusions are feasible in terms of the availability of data related to the economic justification of sustainability.

3. The Result and Discussion

Measurement of efficiency scores using the CCR (Charnes, Cooper, and Rhodes) model through DDEA with input-oriented Constant Return to Scale (CRS) is necessary because the company is in a manufacturing industry that has job order and customization business characteristics, so improving company performance is very dependent on controlling the input of production factors to produce a product.

3.1. Identify the efficiency of sustainability asset indicators based on DDEA

Based on the identification of sustainability asset indicators described earlier, a collective of all aspects of output and input variables, both in economic, social, and environmental aspects, is presented in detail in Table 2.

Table 2. Sustainability asset indicator identification data based on DDEA

DMU Name	DMU No.	Output ∑Product	Input - Economic Indicators			Input - Social Indicator		Input - Environment Indicator			
			Costs of Maintenance (IDR)	Availability (%)	Material Consumption (Pcs/WI)	Environmental Impact (IDR)	Costs of Health (IDR)	Incident & Accident (%)	Material Reject (%)	Hazardous Waste (%)	Energy Consumption (IDR)
180	DMU1	162	26,608	0,996	32.617	882,023	2.085	0,633	0,00183	0,680	5.349
360	DMU2	156	28,457	0,997	52.173	88,924	2.285	0,367	0,02644	0,320	5.529
540	DMU3	197	22,110	0,991	82.832	55,002	2.043	0,429	0,01938	0,453	5.039
720	DMU4	126	1.065,106	0,988	300.623	220,392	1.749	0,571	0,00016	0,547	6.696
900	DMU5	72	565,815	0,996	81.858	633,254	1.034	0,667	0,00001	0,688	5.488
1080	DMU6	174	534,640	0,991	40.111	115,488	1.480	0,333	0,01085	0,312	3.705
1260	DMU7	66	355,961	0,993	28.679	99,157	799	0,143	0,05234	0,659	3.151
1440	DMU8	26	663,647	0,990	40.501	437,368	621	0,857	0,04057	0,341	3.705
1620	DMU9	480	937,323	0,971	161.837	159,252	447	0,200	0,00015	0,580	4.621
1800	DMU10	6	1.171,057	1,000	6.328	0,989	468	0,800	0,00015	0,420	4.722

From the results of data identification in Table 2, data processing was carried out using the Frontier Analyst application to identify the efficiency score of sustainability asset indicators through DDEA. The results of such identification are presented in Table 3.

Table 3. Results of identification of efficiency indicators of sustainability assets based on DDEA

DMU Name	DMU No.	Input-Oriented CRS Efficiency	Sum of lambdas	Return to Scale (RTS)	Optimal Lambdas with Benchmarks	
180	DMU1	1,000	1,000	Constant	1,000	180
360	DMU2	1,000	1,000	Constant	1,000	360
540	DMU3	1,000	1,000	Constant	1,000	540
720	DMU4	0,278	0,263	Increasing	0,263	1620
900	DMU5	1,000	1,000	Constant	1,000	900
1080	DMU6	1,000	1,000	Constant	1,000	1080
1260	DMU7	0,623	0,288	Increasing	0,236	1080
1440	DMU8	0,185	0,098	Increasing	0,069	1080
1620	DMU9	1,000	1,000	Constant	1,000	1620
1800	DMU10	1,000	1,000	Constant	1,000	1800

Table 4. The Results of slack analysis of sustainability asset indicators based on DDEA for 2018 - 2022

DMU Name	DMU No.	Input Slacks								
		Costs of Maintenance (IDR)	Availability (%)	Material Consumption (Pcs/WI)	Environmental Impact (IDR)	Costs of Health (IDR)	Incident & Accident (%)	Material Reject (%)	Hazardous Waste (%)	Energy Consumption (IDR)
180	DMU1	0,000	0,000	0,000	0,000	0,000	0,000	0,000	0,000	0,000
360	DMU2	0,000	0,000	0,000	0,000	0,000	0,000	0,000	0,000	0,000
540	DMU3	0,000	0,000	0,000	0,000	0,000	0,000	0,000	0,000	0,000
720	DMU4	50,726	0,020	41281,306	19,605	369,997	0,107	0,000	0,000	652,535
900	DMU5	0,000	0,000	0,000	0,000	0,000	0,000	0,000	0,000	0,000
1080	DMU6	0,000	0,000	0,000	0,000	0,000	0,000	0,000	0,000	0,000
1260	DMU7	47,005	0,335	0,000	26,278	125,405	0,000	0,030	0,307	849,378
1440	DMU8	58,532	0,086	0,000	68,273	0,000	0,130	0,007	0,025	295,052
1620	DMU9	0,000	0,000	0,000	0,000	0,000	0,000	0,000	0,000	0,000
1800	DMU10	0,000	0,000	0,000	0,000	0,000	0,000	0,000	0,000	0,000

After identifying the efficiency of sustainability asset indicators based on DDEA, slack analysis is carried out on sustainability asset input indicators. Slack analysis of inputs is used to determine the amount of input proportionally so that DMU can reach the minimum point of efficiency. Based on the analysis of slack inputs on sustainability asset indicators, it is known that there are variables with slack input values that have yet to reach the minimum efficiency point target. The results of the slack analysis are presented in Table 4.

Table 5. The Results of the analysis of input targets for sustainability asset indicators based on DDEA for the period 2018 - 2022

DMU Name	DMU No.	Efficiency Input Target								
		Costs of Maintenance (IDR)	Availability (%)	Material Consumption (Pcs/WI)	Environmental Impact (IDR)	Costs of Health (IDR)	Incident & Accident (%)	Material Reject (%)	Hazardous Waste (%)	Energy Consumption (IDR)
180	DMU1	26,608	0,996	32617,000	882,023	2084,961	0,633	0,002	0,680	5348,578
360	DMU2	28,457	0,997	52173,000	88,924	2285,275	0,367	0,026	0,320	5528,969
540	DMU3	22,110	0,991	82831,958	55,002	2043,464	0,429	0,019	0,453	5039,375
720	DMU4	246,047	0,255	42482,213	41,804	117,226	0,053	0,000	0,152	1213,112
900	DMU5	565,815	0,996	81858,000	633,254	1033,950	0,667	0,000	0,688	5488,370
1080	DMU6	534,640	0,991	40111,000	115,488	1479,778	0,333	0,011	0,312	3704,893
1260	DMU7	174,856	0,284	17874,875	35,524	372,339	0,089	0,003	0,104	1114,299
1440	DMU8	64,184	0,097	7489,048	12,602	114,899	0,029	0,001	0,038	390,024
1620	DMU9	937,323	0,971	161837,000	159,252	446,576	0,200	0,000	0,580	4621,378
1800	DMU10	1171,057	1,000	6328,000	0,989	467,665	0,800	0,000	0,420	4722,241

From the slack analysis, it is known that in DMU 4, DMU 7, and DMU 8 there is an excess proportion that has an impact on inefficiencies, so it is necessary to reduce the number of inputs with constant output targets to achieve maximum efficiency targets. The results of the input target analysis are presented in Table 5.

3.2. Analysis of economic aspects and policy implications

The concept is built for companies operating a pool of assets Table high complexity to reduce risk and maximize return on investment must make maintenance strategy a vital issue and invest appropriately in the area [26]. However, based on the results of the analysis of the economic aspects of the efficiency score of sustainability asset performance as in Table 4, it is known that before the organizational change, inefficiencies were found in maintenance costs of IDR 50,726 million with an availability level of 0,993 (less than 100%). In addition, the level of material consumption to produce railway products also experienced inefficiencies of 41.281,306pcs. Compared to the Table of organizational change, the inefficiency of welding machines maintenance costs in 2021 is IDR 47,726 million in semester 1 of 2021 and IDR 58,532 million in semester 2 of 2021. It is also shown that the availability rate of welding machines for production activities throughout 2021 did not reach 100% (semester 1 of 2021 was 0.988, and semester 2 of 2021 was 0.971).

Based on Table 5, to achieve optimal efficiency in maintenance costs before organizational changes in semester 2 of 2019, from the previous IDR 1065,106 million, it must reach the minimum point of IDR 246,047 million at fixed output. Meanwhile, for conditions after organizational changes from the previous IDR 355,961 million (semester 1 of 2021) and IDR 663,647 million (semester 2 of 2021), it must reach the minimum point of IDR 47,005 million and IDR 58,532 million at fixed output.

In addition, from the analysis of target inputs and slack inputs in Tables 4 and 5, it is known that conditions after organizational changes experience inefficiencies that are less optimal than before, followed by differences in management in the asset maintenance function of welding machines.

3.3. Analysis of social aspects and policy implications

Based on Table 4, it is known that during the period of semester 1 of 2018 to semester 1 of 2019, before organizational changes occurred, social aspects had reached optimal efficiency points with a score equal to 1 (one); however, in semester 2 of 2019 before organizational changes, there was excess use of inputs to output with an efficiency score of less than 1 (<1). These advantages are sequentially at the level of environmental impact costs of IDR 19,605 million, health costs of IDR 369,997 million, and the percentage rate of work accidents of 0.107. Meanwhile, after organizational changes throughout 2021, there was excess use of inputs in environmental impact costs of IDR 26,278 million and employee health costs of IDR 125,405 million, and there were still work accidents.

From these conditions, based on Table 5, to achieve optimal efficiency in the period before organizational changes, especially social aspects in semester 2 of 2019, the environmental impact cost from previously amounting to IDR 220,392 million must reach a minimum point of IDR 41,804 million, and employee health costs from the previous IDR 1,749 million per employee must earn a minimum threshold of IDR 117,226 million and an accident percentage rate of 0.053, for fixed output conditions. Meanwhile, to achieve optimal efficiency in the period after organizational changes in semester 1 of 2021, the environmental impact cost from the previous IDR 99,157 million must reach a minimum point of IDR 35,524 million; employee health costs from the previous IDR 799 thousand per employee must achieve a minimum threshold of IDR 372,339 thousand; and the percentage of accident incidence from the last 0.143 must reach a minimum point of 0.089 for fixed output conditions. In Semester 2 of 2021, environmental impact costs from the previous IDR 437,368 million must achieve a minimum threshold of IDR 12,602 million; employee health costs from the previous IDR 621 thousand per employee must reach a minimum point of IDR 114,899 thousand; and the percentage of accident incidence from the last 0.857 must achieve a minimum threshold of 0.029 for fixed output conditions.

3.4. Analysis of environmental aspects and policy implications

Table 4 shows that before the organizational change, environmental aspects were mainly at the optimal point of efficiency with an efficiency score equal to 1 (one), except in the 2nd semester of 2019 when the cost of energy consumption levels exceeded IDR 652,535 million. Meanwhile, conditions after organizational changes throughout 2021 experienced excess use of inputs. In Semester 1 of 2021, the percentage of defective material in the production process (reject) was 0.030, the rate of hazardous waste was 0.307, and the cost of energy consumption was IDR 849.378 million. Meanwhile, for semester 2 of 2021, the percentage of defective material in the production process (reject) is 0.007, the rate of hazardous waste is 0.025, and the cost of energy consumption is IDR 295,052 million.

Based on these conditions, based on Table 5, to achieve optimal efficiency in the period before organizational changes, especially in environmental aspects, the cost of energy consumption level in semester 2 of 2019 from the previous IDR 6,696 million must reach a minimum point of IDR 1,213,112 million. Meanwhile, to achieve the optimal level of efficiency in the period after organizational changes in semester 1 of 2021, the percentage of defective material levels in the production process from the previous 0.052341 must reach a minimum point of 0.003, the rate of hazardous waste from the last 0.659 must achieve a minimum point of 0.104, and the level of energy consumption costs from the previous IDR 3,151 million must reach a minimum threshold of 1,114,299 million. In Semester 2 of 2021, the percentage of defective material levels in the production process from the previous 0.040573 must achieve a minimum point of 0.001. The hazardous waste rate from the last 0.341 must reach a minimum threshold of 0.038. The energy consumption costs from the previous IDR 3,705 million must achieve a minimum point of IDR 390,024 million.

Based on the results of the analysis above, conditions before and after organizational changes, especially in electrical energy consumption, we experienced excess input even though the company already had policies and programs in environmental and energy management. It is followed by the absence of a company program to improve the culture of controlling electrical power for each employee comprehensively, to reduce the company's fixed costs, and provide positive performance in the long run.

4. Conclusion

Based on the results of data processing and data analysis described earlier, it can be concluded as follows:

From the results of the analysis through the Dynamic Data Envelopment Analysis method approach, it is known that the level of efficiency of sustainability asset indicators on the impact of organizational change found in DMU 1, DMU 2, DMU 3, DMU 5, DMU 6, DMU 9, and DMU 10 has shown the achievement of constant or efficient conditions with input-oriented CRS efficiency equal to 1 (DMU = 1). While DMU 4, DMU 7, and DMU 8 show increasing conditions or need to be improved, input-oriented CRS efficiency is at < 1. However, this shows that both before and after organizational changes do not significantly impact the efficiency score of sustainability assets, as seen by the condition of inefficiencies in DMU 4, which was before the organizational change. However, the conditions in question are still better than before the organizational changes.

Based on the results is known that efficiency increases to reach optimal points in DMUs that experience inefficiency conditions, namely:

first, DMU 4 to reach the minimum input point, in maintenance costs, increased efficiency score 77% from Rp1065,106 million, amount of material consumption increased efficiency score 86% from 42482,213pcs/WI, environmental impact costs increased efficiency score 81% from Rp220,392 million, employee health costs increased 93% or Rp117,226 thousand per employee, percentage rate of accidents increased efficiency score 91% from 8 events, dangerous waste level increased efficiency score 72% from 9,570kg and cost energy consumption rate increased efficiency score 82% from Rp1,213,112 million at fixed output.

Second, DMU 7 to reach the minimum input point, the maintenance cost increased efficiency score 51% from Rp174,856 million, the amount of material consumption increased efficiency score 38% from 17,874,875 pcs/wi material, environmental impact cost increased efficiency score 81% from Rp35524 million, employee health costs increased efficiency score 53% from Rp372,339 thousand, percentage accident incidence increased efficiency score 38% from 1 incident, percentage of defective material level in production process increased efficiency score of 94% from 1,584pcs/wi, ratio of hazardous waste increased efficiency score of 84% from 5,102kg and energy consumption cost rate increased efficiency score of 65% from Rp1,114.3 million at fixed output.

Third, DMU 8 to reach the minimum input point, the maintenance cost increased efficiency score by 90% from Rp64,184 million, environmental impact costs increased the efficiency score by 97% from Rp12,602 million, employee health costs increased the efficiency score 81% from Rp114,899 thousand per employee, the percentage of accident incidence increased efficiency score 97% from 6 events, for the rate of defective material levels in the production process increased efficiency score 98% from 1.713pcs/wi, the percentage of hazardous waste increased efficiency score by 89% from 2,643kg and the level of energy consumption cost increased efficiency score by 89% from Rp390,024 million at fixed output.

This research still has things that are lacking to be material for future studies to be refined for future researchers, so it is recommended that, from the condition of organizational changes that have an inefficiency impact on sustainability asset indicators, it is necessary to further evaluate the effectiveness of the company's asset management function, especially in the type of welding machine, and improve the culture in controlling electrical energy for each employee comprehensively, to reduce the company's fixed costs and provide positive performance in the long term. Further research can be carried out in more depth by adding risk analysis to provide mitigation solutions for the impact of corporate organizational change policies. Adding data and research objects to see further comparisons between similar companies will give a better picture of sustainability assets in the existing manufacturing industry.

References

[1] A. Kumaraswamy, R. Garud, and S. (Shaz) Ansari, 'Perspectives on Disruptive Innovations', *Journal of Management Studies*, vol. 55, no. 7, pp. 1025–1042, 2018, doi: 10.1111/joms.12399.

[2] F. Sippl, B. Del Rio, and G. Reinhart, 'Approach for stakeholder identification in Manufacturing Change Management', *Procedia CIRP*, vol. 106, pp. 191–196, 2022, doi: 10.1016/j.procir.2022.02.177.

[3] G. Sætren and K. Laumann, 'Organizational change management theories and safety - A critical review', *10*, no. June 2017, 2017.

[4] Manivelmuralidaran. V, 'Agile Manufacturing - An Overview', *International Journal of Science and Engineering Applications*, vol. 4, no. 3, pp. 156–159, 2015, doi: 10.7753/ijsea0403.1015.

[5] A. Kashikar, D. Mehta, B. Motichandani, and P. D. Chaitanya, 'A case Study on Agile And Lean Project Management In Construction Industry', *IOSR Journal of Mechanical and Civil Engineering*, vol. 13, no. 04, pp. 31–39, 2016, doi: 10.9790/1684-1304013139.

[6] M. of S.-O. Enterprises, 'Kementerian Badan Usaha Milik Negara'. Accessed: Nov. 13, 2021. [Online]. Available: https://bumn.go.id/portfolio

[7] E. Cagno, A. Neri, M. Howard, G. Brenna, and A. Trianni, 'Industrial sustainability performance measurement systems: A novel framework', *J Clean Prod*, vol. 230, pp. 1354–1375, 2019, doi: 10.1016/j.jclepro.2019.05.021.

[8] M. Ghaleb and S. Taghipour, 'Assessing the impact of maintenance practices on asset's sustainability', *Reliab Eng Syst Saf*, vol. 228, no. December 2021, p. 108810, 2022, doi: 10.1016/j.ress.2022.108810.

[9] J. P. Liyanage, F. Badurdeen, and R. M. C. Ratnayake, 'Industrial Asset Maintenance and Sustainability Performance : Economical , Environmental , and Societal Implications', *Industrial Asset Maintenance and Sustainability Performance*, pp. 665–693, 2009, [Online]. Available: https://doi.org/10.1007/978-1-84882-472-0_24

[10] K. C. Sinha, S. Labi, and B. R. D. K. Agbelie, 'Transportation infrastructure asset management in the new millennium: continuing issues, and emerging challenges and opportunities', *Transportmetrica A: Transport Science*, vol. 13, no. 7, pp. 591–606, 2017, doi: 10.1080/23249935.2017.1308977.

[11] A. Abdi and S. Taghipour, 'Sustainable asset management: A repair-replacement decision model considering environmental impacts, maintenance quality, and risk', *Comput Ind Eng*, vol. 136, no. July, pp. 117–134, 2019, doi: 10.1016/j.cie.2019.07.021.

[12] PT Industri Kereta Api, *Rencana Kerja dan Anggaran Perusahaan*, 2021st ed. PT Industri Kereta Api, 2021.

[13] N. Almeida, M. Trindade, D. Komljenovic, and M. Finger, 'A conceptual construct on value for infrastructure asset management', *Util Policy*, vol. 75, no. October 2019, p. 101354, 2022, doi: 10.1016/j.jup.2022.101354.

[14] D. Maletič, N. Marques de Almeida, B. Gomišček, and M. Maletič, 'Understanding motives for and barriers to implementing asset management system: an empirical study for engineered physical assets', *Production Planning and Control*, vol. 0, no. 0, pp. 1–16, 2022, doi: 10.1080/09537287.2022.2026672.

[15] A. Trianni, E. Cagno, A. Neri, and M. Howard, 'Measuring industrial sustainability performance: Empirical evidence from Italian and German manufacturing small and medium enterprises', *J Clean Prod*, vol. 229, pp. 1355–1376, 2019, doi: 10.1016/j.jclepro.2019.05.076.

[16] I. H. Garbie, 'An analytical technique to model and assess sustainable development index in manufacturing enterprises', *Int J Prod Res*, vol. 52, no. 16, pp. 4876–4915, 2014, doi: 10.1080/00207543.2014.893066.

[17] Y. C. Wu and S. W. Lin, 'Efficiency evaluation of Asia's cultural tourism using a dynamic DEA approach', *Socioecon Plann Sci*, vol. 84, no. July, p. 101426, 2022, doi: 10.1016/j.seps.2022.101426.

[18] N. Amowine, Z. Ma, M. Li, Z. Zhou, B. A. Asunka, and J. Amowine, 'Energy efficiency improvement assessment in Africa: An integrated dynamic DEA approach', *Energies (Basel)*, vol. 12, no. 20, 2019, doi: 10.3390/en12203915.

[19] H. Nedaei, S. G. Jalali Naini, and A. Makui, 'A dynamic DEA model to measure the learning rates of efficient frontier and DMUs: An application to oil and gas wells drilling', *Comput Ind Eng*, vol. 144, no. February 2019, p. 106434, 2020, doi: 10.1016/j.cie.2020.106434.

[20] H. Omrani and E. Soltanzadeh, 'Dynamic DEA models with network structure: An application for Iranian airlines', *J Air Transp Manag*, vol. 57, pp. 52–61, 2016, doi: 10.1016/j.jairtraman.2016.07.014.

[21] K. Tone and M. Tsutsui, 'Dynamic DEA with network structure: A slacks-based measure approach', *Omega (United Kingdom)*, vol. 42, no. 1, pp. 124–131, 2014, doi: 10.1016/j.omega.2013.04.002.

[22] K. Tone and M. Tsutsui, 'Dynamic DEA: A slacks-based measure approach', *Omega (Westport)*, vol. 38, no. 3–4, pp. 145–156, 2010, doi: 10.1016/j.omega.2009.07.003.

[23] V. J. Bolós, R. Benítez, and V. Coll-Serrano, 'Continuous models combining slacks-based measures of efficiency and super-efficiency', *Cent Eur J Oper Res*, vol. 31, no. 2, pp. 363–391, 2023, doi: 10.1007/s10100-022-00813-5.

[24] A. Charnes, W. W. Cooper, and E. Rhodes, 'Measuring the efficiency of decision making units', *Eur J Oper Res*, vol. 2, no. 6, pp. 429–444, 1978, doi: 10.1016/0377-2217(78)90138-8.

[25] I. H. . Elkington, John; Rowlands, 'Cannibals with forks: the triple bottom line of 21st century business', *Alternatives Journal Waterloo*, vol. 25, no. 04, pp. 42–43, 1999, doi: 10.5860/choice.36-3997.

[26] W. N. Cahyo, *Engineering Asset Management(Pengantar Manajemen Aset Industri berbasis ISO 55000)*. Daerah Istimewa Yogyakarta: Universitas Islam Indonesia (Islamic University of Indonesia), 2019.

IOP Conf. Series: Earth and Environmental Science 1256 (2023) 012002 doi:10.1088/1755-1315/1256/1/012002

Integrated 3R Waste Processing Facility in Batan Indah as an Educational Waste Tourism Center

Irviananda Adenia, Ratna Safitri, Melania Lidwina Pandiangan

Department of Architecture, Universitas Pembangunan Jaya, Tangerang Selatan, Indonesia

irviananda.adenia@student.upj.ac.id

Abstract. The management of waste has emerged as a pressing global concern due to its profound environmental and societal consequences. This paper presents the design proposal for an integrated waste processing facility, named Batan Indah, which serves as both a solution to waste management challenges and an educational center for waste awareness and education. The facility aims to exemplify the principles of Reduce, Reuse, and Recycle (3R) while offering visitors an immersive learning experience. Architecturally, the facility blends harmoniously with its natural surroundings, showcasing a strong connection to the environment. Abundant natural lighting and ventilation minimize dependence on artificial energy sources. The design proposal for the integrated waste processing facility in Batan Indah presents an innovative approach to waste management, emphasizing its educational aspect. By combining sustainable architectural design, immersive learning experiences, and community engagement, the Batan Indah facility seeks to promote the principles of the 3R concept and inspire visitors to actively participate in waste reduction, reuse, and recycling initiatives.

1. Introduction

The increasing population of South Tangerang City each year, coupled with changes in consumer patterns, has led to numerous environmental impacts, particularly in the realm of waste. Consequently, there has been a surge in the volume and diverse characteristics of accumulating waste. Recognizing this upward trend, the local government of South Tangerang City has implemented waste management policies through the 3R Integrated Waste Processing Facility/*Tempat Pembuangan Sampah Terpadu* (TPST) program, which emphasizes Recycle, Reuse, and Reduce practices [1]. This program is outlined in Regional Regulation No. 3 of 2013. The core concept of the TPST 3R program revolves around improving the quality and characteristics of waste destined for final disposal sites/*Tempat Pembuangan Akhir* (TPA) while simultaneously reducing the quantity of waste entering the system.

Insufficient waste management practices, particularly in developing countries, can have negative impacts on public health in surrounding communities [2]. A tangible example of this issue is the defunct Cipeucang Final Disposal Site (TPA) in South Tangerang. This TPA has been deemed environmentally polluting due to exceeding waste capacity, resulting in excessive landfill mounds. To alleviate this problem, the TPA Cipeucang has to transfer more than 20 trucks of waste daily to the Serang TPA. To prevent a recurrence of this situation, South Tangerang requires more adequate Integrated Waste Processing Facilities (TPST) to increase the amount of recyclable waste processed before reaching the

Cipeucang TPA. One of the largest TPST facilities in the South Tangerang area with the potential to process a significant amount of waste prior to disposal at the Cipeucang TPA is TPST 3R Batan Indah.

The limited recycling activities and programs currently in place align with the low level of implementation and understanding of the 3R principles (Reuse, Reduce, Recycle) among Indonesian society [3]. Considering the significant waste generated and the lack of recycling and waste segregation practices by the population, waste management issues persist. Without immediate attention and socialization of a new waste management approach that emphasizes waste reduction and recycling from collection to disposal, the waste will not be properly managed, resulting in suboptimal functionality of TPAs and TPSTs in the South Tangerang region.

Community behavior also needs to be educated to minimize waste accumulation in South Tangerang. In addition to the lack of education, this behavioral trend is also influenced by the absence of communication and transparency regarding waste management practices, leading to a lack of public awareness regarding the urgency of waste issues in the region. Education plays a crucial role in influencing environmental behavior in waste management [4]. The lack of education results in limited public understanding of waste management processes. Therefore, transparency supported by education is crucial, especially considering that the TPST 3R facility is currently concealed and hidden.

Therefore, there is a need for further design development of the TPST 3R Batan Indah to enable the proper and efficient processing of waste on a larger scale while also serving as an educational tourist attraction. The education provided should focus on waste segregation and various waste processing techniques, ranging from simple to complex methods. The concept of TPST 3R as an educational tourist attraction should be incorporated into the architectural design to ensure transparency, allowing both informed and uninformed visitors to witness and become aware of the waste issues in our surroundings.

This combined educational tourist facility and TPST 3R function will be utilized to process selected waste while also serving as an educational and tourism destination for the South Tangerang community. Waste items that can be processed, particularly at TPST 3R Batan Indah, such as household waste, plastic bottles, coconut shells, glass, plastic packaging, and cardboard, can undergo transformation into valuable products. For instance, they can be turned into eco-bricks, explore various ceramic motifs, furniture, and other construction materials.

The purpose of this article is to elaborate the TPST 3R Batan Indah design that introduces an innovative waste management and education approach through transparency, aiming to raise public awareness and educate about waste generation. This design is a final project of an architecture study program.

2. Methodology

In this context, TPST 3R Batan Indah, located in Kademangan, specifically in the Setu District, has successfully accommodated the collection and processing of household waste for all residents of the Batan Indah Complex, consisting of 1100 households (KK). TPST 3R Batan Indah is capable of transporting 2-2.5 tons of waste daily. With a total of 12 employees, TPST 3R Batan Indah categorizes the waste into two main types: organic and non-organic waste. In the waste sorting process, TPST 3R Batan Indah further classifies organic waste into two categories: sellable waste items, such as plastic bottles, plastic cups, cardboard, and glass; and non-sellable waste items, such as plastic packaging from coffee and leftover food containers, which need to be shredded and processed to create marketable products. For organic waste, TPST 3R Batan Indah processes coconut shells and household waste using maggots to produce compost and animal feed.

This study presents an exploratory design study that addresses an existing problem. In this case, the problem to be addressed is the importance of developing the design of TPST 3R Batan Indah as a new waste educational tourism destination. This involves the need to develop the design of TPST facilities that can accommodate the increasing volume and types of waste while educating the community about the 3R principles through the application of transparent architectural concepts.

3. Literature Review

3.1. Waste Management

Waste management is an activity aimed at reducing and handling the waste generated by living organisms. Waste management activities can include collection, transportation, processing, recycling, or disposal of waste materials [5].

Integrated Waste Processing Facilities (TPST) are places where waste collection, sorting, reuse, recycling, and processing activities take place [6]. In the waste management hierarchy, there are three main methods: material reuse, composting, and waste reduction, commonly known as the 3R approach, which emphasizes the need to reduce, reuse, and recycle waste to create value [7].

TPST 3R also has specific facility standards that must be provided to facilitate the waste processing process. Facilities in the Integrated Waste Processing Facilities include pre-processing facilities, sorting facilities, physical waste processing facilities, and advanced processing facilities [1].

3.2. Educational Tourism and Transparency

To attract the attention of visitors and enhance waste management awareness, one approach is to combine the TPST with a tourism element, creating an educational tourism experience. Developing educational tourism involves several considerations, such as basing it on learning activities, increasing visitors' knowledge, incorporating unique elements, and designing it in the form of an adventure or specific narrative [8].

The success of education can be supported by transparency. Transparency in architecture refers to the use of open materials in building envelopes, allowing for visibility of the activities taking place inside. Transparency in architecture is used to depict conditions that allow someone to see from the outside into the interior space, and vice versa [9]. This concept aims to connect the visual aspects of the interior and exterior spaces. Additionally, transparency can also be achieved by creating partially transparent spaces, where the state cannot be visibly observed due to the use of materials and partition boundaries, but can be felt [10]. Examples of implementation include creating differences in leveling and the size of the spaces. Transparency can also be achieved through translucency or the use of transparent materials.

3.3. Design Criteria

The location for constructing TPST 3R should ideally be within flood-free residential areas, with its own access road but not too far from the main road. This is crucial for the convenience of waste collection trucks. The placement of TPST 3R should be as close as possible to the serviced areas within a radius of no more than 1 km. The location and visitor capacity must meet the needs and should be easily accessible without causing environmental pollution. TPST 3R should also have a fixed schedule for waste collection and transportation from residents.

Based on the understanding of waste management, educational tourism concepts, and transparency in architecture, the design should meet the criteria outlined in Table 1.

Table 1. Design Criteria

	Design Components	Transparency in Architecture	Design Criteria
Waste Processing Space	Public	Visual connection between exterior and interior spaces	Existence of connectivity between outdoor and indoor spaces through an open building envelope
	Semi-Public	Interiors without column restrictions or with translucent materials to enable visibility of	Open column-free spaces within the educational tourism area to enhance

		activities throughout the space	visual engagement during learning activities
	Private	Non-transparent elements used for interconnecting spaces	Private spaces devoid of transparency in architectural design
Material	Utilization of recycled materials in the TPST	Use of green and recycled materials that are open for both interior and exterior applications	Incorporation of green materials and recycled waste products in the open and translucent elements of the TPST's exterior and interior.

4. Results and Discussions

4.1. Facility Design

The Integrated Waste Treatment Facility (TPST) 3R Batan Indah is located in Setu District, South Tangerang City. The site is situated along Serpong Highway, specifically within the Batan Indah Complex. The total area designated for TPST development is 15,000m2 or 1.5 hectares, and the currently constructed TPST 3R occupies an area of 3,394m2. The northern and western parts of the site are adjacent to the Batan Indah Complex, the eastern part borders Serpong Highway, and the southern part borders the Thousand Heroes Cemetery Park. Access to the site can be achieved by private vehicles such as cars and motorcycles, given the wide road, which also allows the passage of waste collection trucks. For public transportation, there is the Serpong Commuter Line (KRL) Station located 2 km away from the TPST. In this study, the boundaries of the new TPST 3R Batan Indah can be seen in Figure 1, including the existing TPST, undeveloped areas, and a brownfield area to establish a connection towards the park. Opening a connection towards the park aims to expand the public area and generate interest and attraction for visitors to the waste management education center.

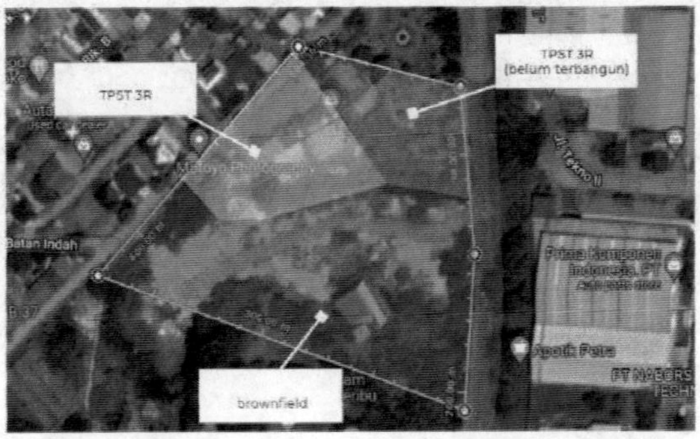

Figure 1. Development boundary plan of TPST 3R Batan Indah.

Each area in TPST 3R Batan Indah (Figure 2) serves multiple functions and is used depending on its schedule. For example, the area used for waste sorting also serves as a parking area for waste bins and armroll containers in the afternoon after the sorting process is completed. The maggot cultivation area also has a relatively small cultivation space, so if there is an excess of non-organic waste, the outdoor area next to the cultivation space will be used.

Figure 2. Existing facilities area at TPST Batan Indah.

In its daily practice, TPST 3R Batan Indah follows a waste management process from collection to processing. Initially, waste collectors will collect waste from households in Komplek Batan Indah every morning from Monday to Friday, completing the collection by 11:00 AM local time. After that, TPST personnel will start sorting the waste into three categories: organic waste, non-organic waste, and electronic waste. Electronic waste is separated in a dedicated room, while organic waste is piled up in an arm roll container. Non-organic waste is sorted using a conveyor system to separate it based on its type. TPST 3R Batan Indah further divides non-organic waste into five categories: plastic bottles, plastic glasses, cardboard, glass, and packaging waste leftovers.

After sorting with the conveyor, sellable waste is weighed using a weighing machine and placed in selling bags. Sellable waste includes plastic bottles, plastic glasses, cardboard, and glass. However, there is waste that doesn't have any commercial value, such as leftover plastic packaging waste. These packaging waste leftovers need to be processed further, including shredding, to be transformed into items with commercial value such as furniture, chairs, tables, building materials, and others.

For the processing of organic waste, TPST 3R Batan Indah divides it into two types: household food waste and coconut husks. Household food waste is turned into compost through maggot processing methods. This organic waste can also be used as animal feed for ducks and goats, which are also raised at TPST 3R Batan Indah. As for coconut husks, they are processed and shredded to become cocopeat, a high-water absorption planting medium.

4.1.1. Pre-processing facility (Figure 3). After the waste is collected from residential areas, the waste bins are parked in the designated area of TPST 3R Batan Indah, which is located near the temporary waste storage area before sorting.

4.1.2. Sorting facility (Figure 4). Once all the waste from households is gathered in the temporary storage area, it is manually sorted into organic and non-organic waste. The organic waste is placed in an arm roll container. The non-organic waste, specifically, goes through a conveyor system for further sorting. The conveyor helps separate the waste that can be sold (such as bottles, plastic, glass, and cardboard) from the waste that needs further processing, such as food plastic waste. The sellable waste is weighed and packed into selling bags. These bags are stored in both indoor and outdoor storage areas. The indoor space is used for storing glass, cardboard, and plastic glasses, while the outdoor area is used to store the sorted plastic bottles and plastic glasses before they are sold.

4.1.3. Coconut husk processing facility (Figure 5). Coconut husks at TPST 3R Batan Indah can be processed into a planting medium called cocopeat. Each month, this facility can process 2 tons of coconut husks. The coconut husks are first collected in a designated area to prevent mixing with other waste before being fed into a grinding machine. The coconut husks are then crushed using a disk mill

IOP Conf. Series: Earth and Environmental Science 1256 (2023) 012002 doi:10.1088/1755-1315/1256/1/012002

machine, transforming them into fibers. These fibers are further refined through a coconut screening machine, which rotates to break them down into smaller pieces resembling soil or planting medium. The finely processed coconut husks are stored in containers and left to mature until they turn brown. The brown color indicates that the coconut husks are ready to be packaged and used as cocopeat planting medium.

4.1.4. Food waste processing facility (Figure 6). The sorted food waste at TPST 3R Batan Indah is processed into animal feed and fertilizer using maggots. Maggots, also known as larvae or fly larvae, are mixed with the waste in a container box and undergo five stages of maggot development before becoming animal feed and fertilizer. After transferring the food waste to the container with maggots, the containers are arranged and moved to a designated area that is shielded from direct sunlight.

4.1.5. Plastic packaging waste processing facility (Figure 7). Plastic waste that cannot be sold directly undergoes shredding to be transformed into valuable products. The shredded plastic waste can be used for various purposes, such as furniture (tables and chairs), decorative wall coverings, and wall or floor coverings (ceramics). At TPST 3R Batan Indah, the plastic packaging waste is washed and cleaned before being placed in large bags for transportation to the Refused Derived Fuel (RDF) Integrated Waste Treatment Facility in Cilacap for shredding. Shredding cannot be performed at the TPST 3R Batan Indah due to limited space, preventing the shredding machine from entering the processing area. The shredded waste is returned to TPST 3R Batan Indah to be processed into other products. However, due to land and equipment limitations, the shredded waste is buried and not further processed.

Figure 3. Pre-processing facility.

Figure 4. Sorting facility.

Figure 5. Coconut shell waste processing facility.

Figure 6. Food waste processing facility.

Figure 7. Plastic packaging waste processing facility.

4.2. Transparency as Waste Processing Educational Tourism
Based on the identified issues and the desired objectives of the TPST 3R Batan Indah design development, the theme of this design focuses on waste education tourism that emphasizes transparency in architecture. The design incorporates three main principles: expose, elevate, and enliven, which are integrated into the circulation and visitor experience. This theme is for the development plans of the landfill in the future, to maximize the remaining land that has not been processed by the TPST 3R Batan Indah, as a waste education tourism center.

4.2.1. Expose. The first phase aims to raise public awareness about the significant volume, types, and characteristics of waste generated by human activities on a daily basis. This phase explains and visually presents the waste management process, different types of waste, and the waste sorting procedures implemented at TPST 3R Batan Indah. The intended learning outcome in this phase is for waste education tourists to recall and comprehend the various types and characteristics of waste.

4.2.2. Elevate. In the second phase, visitors will observe how the sorted waste is processed into new products with added value, benefiting the surrounding environment. This phase explains the direct recycling processes of coconut husks, food waste, and plastic waste. To enhance the understanding of the processing procedures, visitors can also observe the production flow through animations presented in a theater space. The expected learning outcome in this phase is for waste education tourists to analyze the ongoing waste processing procedures and relate them to the information presented in the previous phase.

4.2.3. Enliven. During the enliven phase, visitors are encouraged to contribute to waste management by transforming existing waste into new products with commercial value. This phase aims to stimulate creativity and raise visitors' awareness to actively participate in maintaining a comfortable living environment for the future. In the enliven area, visitors will directly witness the processed waste, including interactive installations, as well as the use of compost and animal feed to green the open green spaces of TPST 3R Batan Indah. The expected learning outcome in this phase is for waste education tourists to apply, create, and evaluate the waste management process, starting from waste sorting to reuse for daily needs.

To achieve the three main principles of the TPST 3R Batan Indah design, the site area will be divided into three sections corresponding to each phase that visitors will experience. The expose phase, located after the entrance, aims to reveal and educate visitors about the actual urgency of waste management, particularly in the city of Tangerang Selatan. The elevate phase will serve as the area showcasing the waste processing processes, including coconut husks, food waste, and plastic packaging. Adjacent to the second phase area, there will be a control panel room and a warehouse for operating and storing heavy equipment. In the final enliven phase, visitors can directly contribute to recycling the existing waste at TPST 3R Batan Indah. This area will also demonstrate the outcomes of processed waste and their benefits for the environment.

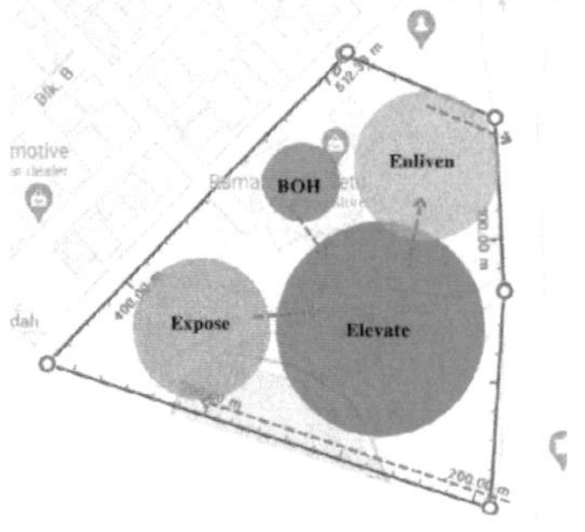

Figure 8. Zoning of the three supporting phases in the transparency concept.

The development of the TPST 3R Batan Indah design will also involve the construction of a 3-storey building (Figure 9). The majority of the first floor will be dedicated to waste processing, from sorting to value-added processes. On the first floor, there will be visitor parking areas, waste truck parking areas, waste sorting and weighing areas, waste sales areas, waste processing areas, a playground area, green open spaces, equipment storage, a control panel room, and installation areas. The second floor will primarily focus on the educational tourism objectives, featuring a food court, shops selling products made from recycled waste for visitors to purchase, a theater room, and a multipurpose room. The third floor will accommodate office and administrative spaces, as well as rest areas for workers and employees.

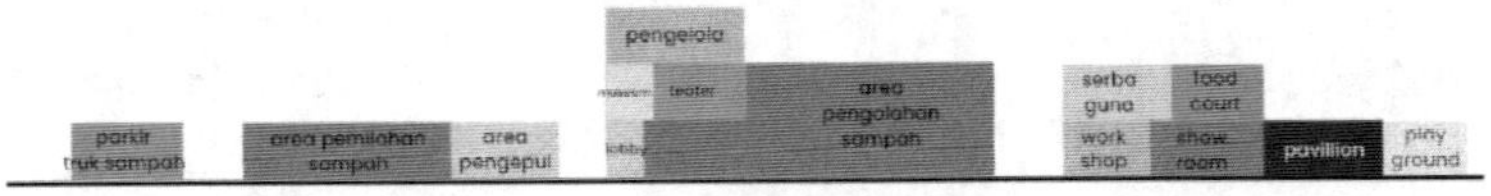

Figure 9. Vertical zoning.

The TPST 3R Batan Indah design will also incorporate interior and exterior concepts using recycled materials directly obtained from the waste processing. The façade will incorporate recycled plastic and coconut husk layers in certain sections to directly showcase the processed waste to the waste education tourists. The interior will also feature recycled waste materials, visible in the ceiling, wall coverings, flooring, and furniture such as tables and chairs made from waste materials. To emphasize the concept of transparency, the exterior and interior spaces will be connected through a facade with open and light-transmitting materials. The relationship between the interior spaces will be column-free, allowing a view of the entire waste processing activity within the same area.

4.3. TPST Design Outcome

Based on the literature review and the formulation of design criteria, the TPST 3R Batan Indah design outcome resembles a waste mountain shape (Figure 10). The building adopts a dome system divided into three parts corresponding to the transparency phases. On the ground floor, the first building houses the waste processing facilities, the second building accommodates waste sorting activities, and the third building serves as a commercial area, such as a museum and display area for visitors (Figure 11). The site is not enclosed by fences to enhance transparency from the outside to the site area, and visitors can access it from the nearby park. Waste trucks and visitors have separate circulation routes, with visitors entering from the main road and waste trucks entering from the rear area. The waste distribution process from the waste collection area to sorting and processing areas utilizes conveyors to facilitate the transfer of waste and make the process faster and more hygienic.

Figure 10. Resembling a pile of waste.

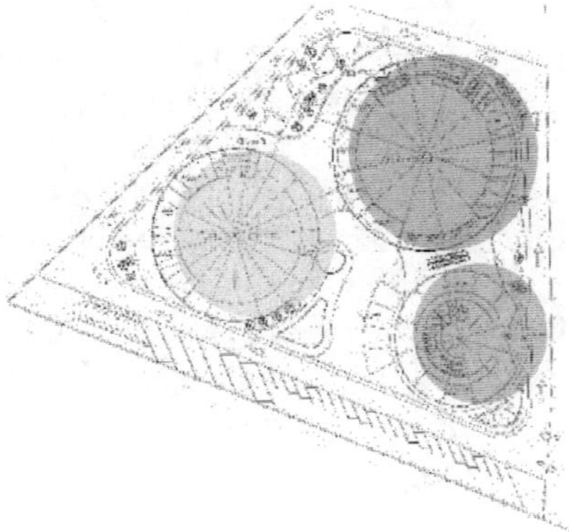

Figure 11. Zoning on the ground floor.

Visitors can observe the entire waste processing and sorting process through walkways on the second and third floors. In addition to accessing the walkways from the main lobby, visitors can directly access them from the outdoor areas adjacent to the park (Figure 12). This is expected to attract a wider audience interested in observing and learning about the waste processing process. Increased engagement will lead to greater awareness of the importance of environmental care, including waste separation and processing.

Figure 12. Exterior walkway.

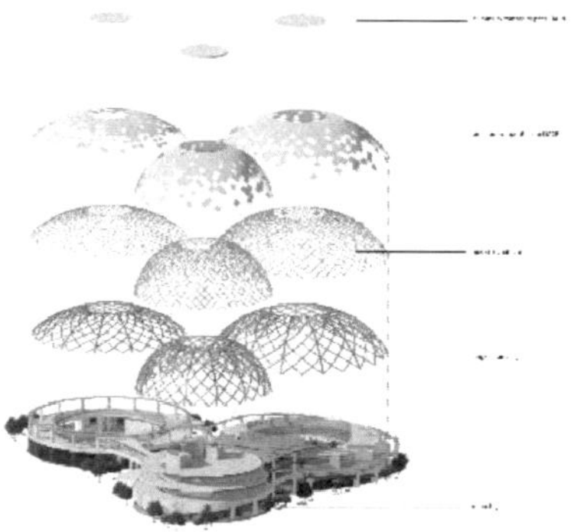

Figure 13. Combination of solid and transparent materials in the roof.

Furthermore, the dome-shaped building's roof covering material combines transparent and solid materials (Figure 13). Transparent materials align with the increased transparency, allowing the public to observe the waste processing process from the walkways surrounding the façade. It also enhances the utilization of natural light within the building. However, it is important to balance this with the use of solid materials to prevent excessive light from entering. Excessive light and heat can reduce the comfort level for building users.

5. Conclusions

In conclusion, the design of the TPST 3R Batan Indah presents an innovative approach to waste management and education. By incorporating transparency as a guiding principle, the design aims to raise awareness and educate the public about the volume, types, and characteristics of waste generated in daily activities. The three phases of the design, namely Expose, Elevate, and Enliven, provide visitors with a comprehensive understanding of waste processing, from sorting to value-added transformations.

This concept holds novelty value in the field of waste management and sustainable tourism. The integration of recycled materials throughout the interior and exterior emphasizes the concept of sustainability and serves as a tangible demonstration of waste transformation. The use of transparent and solid materials in the façade and roof allows for natural light while maintaining a balance to ensure occupant comfort. The TPST 3R Batan Indah design promotes engagement and interaction by providing walkways and viewing areas that allow visitors to witness the waste processing and sorting processes. The design also facilitates public access and integrates the site with the surrounding environment, creating a seamless connection between the interior and exterior spaces.

Overall, the TPST 3R Batan Indah design presents a holistic and engaging approach to waste management education, offering a unique opportunity for visitors to learn about waste processing, contribute to recycling efforts, and cultivate a deeper understanding of environmental stewardship. This design exemplifies the potential for integrating sustainable principles and educational initiatives to address the pressing issue of waste management in an aesthetically appealing and informative manner.

6. References

[1] Roos, N., A, A., & S, N. (2022). KAJIAN PERENCANAAN TEMPAT PENGOLAHAN SAMPAH TERPADU DI DESA KARANGKATES KABUPATEN MALANG. *Prosiding SEMSINA*, *3*(2). https://doi.org/10.36040/semsina.v3i2.5139

[2] Wilson, D. C., Velis, C. A., & Rodic, L. (2013). Integrated sustainable waste management in developing countries. *Proceedings of Institution of Civil Engineers: Waste and Resource Management*, *166*(2). https://doi.org/10.1680/warm.12.00005

[3] Sujadmi, Hayati, L., & Saputri, R. A. (2021). Empowering Society in Waste Management System with the Reduce Reuse and Recycle Approach in Pagarawan Bangka. *IOP Conference Series: Earth and Environmental Science*, *926*(1). https://doi.org/10.1088/1755-1315/926/1/012020

[4] Ridayani, Saputra, N., Siagian, N., Owon, R. A. S., & Rawadhy, I. (2022). The correlation of environmental education, environmental knowledge, environmental involvement, and waste management behavior. *IOP Conference Series: Earth and Environmental Science*, *1105*(1). https://doi.org/10.1088/1755-1315/1105/1/012008

[5] Nafiah, C. F., Fadilah, K., & Lukita, C. W. (2023). Perencanaan Detail Engineering Design (DED) Tempat Pengolahan Sampah Terpadu (TPST) Di Tempat Pembuangan Akhir (TPA) Regional Banjarbakula Kalimantan Selatan. *Jurnal Teknik Sipil Dan Lingkungan*, *8*(1), 37–46. https://doi.org/10.29244/jsil.8.1.37-46

[6] Sukwika, T., Sukamdani, N. B., & Maharani, M. D. D. (2022). Analysis of Two Dimensional Sensitivity as Decision Support Alternative Waste Management in TPST Bantargebang. *BERKALA SAINSTEK*, *10*(2). https://doi.org/10.19184/bst.v10i2.31293

[7] Sutanto, H. B. (2021). Pengembangan Wisata Edukasi Sampah Berbasis Komunitas di Kelurahan Sorosutan, Yogyakarta. *Sendimas 2021 - Seminar Nasional Pengabdian Kepada Masyarakat*, *6*(1). https://doi.org/10.21460/sendimasvi2021.v6i1.71

[8] Zamzani, M. I., Malik, A. R., Ayu, D., Sukma, F., Raga, G., & Alik, M. R. (2022). Meningkatkan Kawasan Bendali II menjadi Tempat Wisata Edukasi Berwawasan Lingkungan. *Jurnal Inovasi Pengabdian Dan Pemberdayaan Masyarakat*, *2*(2). https://doi.org/10.54082/jippm.46

[9] Islami, S. Y. (2011). The Opacity of Glass Rethinking Transparency in Contemporary Architecture. *International Journal of Architecture and Urban Development*, *1*(2). https://www.researchgate.net/publication/361342267

[10] Shahlaei, A., & Mohajeri, M. (2015). In-Between Space, Dialectic of Inside and Outside in Architecture. *INTERNATIONAL JOURNAL OF ARCHITECTURE AND URBAN DEVELOPMENT*, *5*(3).

7. Acknowledgments

The author would like to express gratitude to the Architecture Study Program at Pembangunan Jaya University, South Tangerang, for their sponsorship of this activity. The author would also like to thank TPST 3R Batan Indah for providing data and information throughout the course of this activity. This research is part of the Final Project of the Architecture Study Program at Pembangunan Jaya University.

Residential's Household Intention to Utilize Their Food : A Case Study in Solo, Indonesia

Raden Danang Aryo Putro Satriyono, Dinda Safitri Ramadani
Faculty of Engineering, Universitas Muhammadiyah Surakarta

Rda715@ums.ac.id

Abstract. Food loss and food waste leads to severe effect such as economic, environment and social consequences in Indonesia as the third largest contributor of food waste In the world. Most of the food waste source in Indonesia is from household sector. This phenomenon needs to be analyzed more deeply with tested variables related to human behaviour and perception. This study investigates determinants of residents' participation intention and behaviour to waste their food in a framework that incorporates Extended theory of planned behaviour (TPB). The modified TPB questionnaire was given to 300 respondent samples ever used waste bank in Solo, Central Java. The analysis in this study was carried out using Structural Equation Modeling (SEM), The result of this research is that there are five significant correlations between variables, namely Perceived Circular Knowledge, Perceived Economic Usage, Perceived Economic Knowledge, Descriptive Norms and attitude have a significant effect on Intention to Use. This study also found that the Perceived Effectiveness, Perceived Usefulness, Perceived Behavioural Control factors had no significant effect on Intention to Use.

1. Introduction

Indonesia, a massive country with over 275 million people in it was combating with food waste emergency. This phenomenon alarmed the ministry of indonesian forest and environment who estimated that average food waste to be 0.07 kg per person (8% of the total waste rate, or 0.87 kg per person per day) (Jambeck et al, 2015). According to this information, food waste amounted 7 million tons were generated, of which 4.9 million tons were claimed to have been improperly managed and released into the natural reservoir (Ren et al, 2023). Being listed as one of the largest producer of food waste worldwide (Surya et al, 2020). Policy maker substantially increase the awareness regarding waste pollution by applying regional waste management strategies with the goal of reducing solid waste at source by 30% by 2025, in accordance with Presidential Decree No. 97/2017 on National Policies and Management Strategies for Domestic and Similar Wastes (Jakstrada) (Merli et al, 2018). Other thing to address is about our single-use plastics waste (shopping bags, straws, polystyrene wrappers, etc.). One initiative to address this issue is the introduction of a circular economy.

The circular economy is defined as an system of economic that holds materials and goods in use for as long as feasible in order to maximize used resource and minimize waste (Hettich et al, 2021). Circular economy have three disctintive characteristic which are try to do restoration rather than efficiency of waste, engineered out sustainability concept from waste and pollution and try to eliminate waste and pollution as much as possible (Garcia-Muina et al, 2018). Another circular economy principle is keeping products and materials in use through repair, reuse and recycling (Miftahorrozi et al, 2022). Restoring and regenerating natural systems to ensure that next generations will have chance to benefitted from their resources, such as woods, energy, streams, and soil (Bagastyo et al, 2022).

The circular economy seeks to replace the linear "take, make, dispose" model with a "reduce, reuse, recycle" framework that is more environmentally friendly. Thanks to the circular economy, the environmental impact is reduced and there are opportunities for people to be creative and innovative to boost economic development (Hellmich et al, 2019). In the past few years, the notion of economic circular has surfaced as a sustainability blueprint, the ability to regenerate itself, and the reevaluation of conventional production models that rely on excessive use of resources and the discarding of used items (Zhang et al, 2020). This concept rethinking the entire chain of supply from concept to production, use, disposal and innovations and technologies that turn waste into resources. The circular economy means that company enables them to become more resilient and competitive to achieve their goals that have a positive impact on the environment (Hadibasyir et al, 2020). Nonetheless, shifting to circular economy requires a cooperation among educational and research institution, industry, governments and consumer. The economic circular is diverse and involves all stakeholders. We need a shared vision for circular economic transition and need to bring stakeholders together to develop transformational strategies (Urakami et al, 2021). The rapidly evolving of circular economic into a model of sustainability and a self-renewable economy (Chakaborty et al, 2018). Industrialization has improved people's living standards, but it has also destroyed the environment through overconsumption and overproduction. The circular economy is based on two main aspects, both biologically and technically. The goal of the circular economy is to minimize polluting waste by preserving materials in closed-loop systems for as long as possible. Aspects of biology and technology play a major role in this system in enabling more resource-efficient use of those resources.

Despite regulatory approaches, community engagement to reduce the amount of food waste is essential to solve the problem. Determining the extent of community knowledge, attitudes, perceptions and engagement is critical to effectively implementing a food waste reduction program such as circular economy. Although there have been some extensive studies of public attitudes, behaviors and practices towards food waste. In fact, there is still a certain information gap in Indonesian society regarding attitudes and behaviour towards food waste disposal. As per 2012, Indonesia has a program called waste bank as a part of national solid waste management strategies in a number of TPS 3R (Gumasing et al, 2021). The presence of waste banks may have improved socioeconomic empowerment, perspective, and community awareness of solid waste management (Szalavetz et al, 2019). This study then investigated attitude, perception, household knowledge about food waste, reason to waste their food and participation level on implementing circular economy and management to tackle food waste problem.

2. Research methods

One of the behavioral model theories commonly used in behaviour and perception-related research is Theory of Planned Behaviour (TPB). TPB has been used specifically to analyze perception and habits in various studies around the world (Nittono et al, 2020). Related to circular economy, Research by Zhang et al (2010) has developed the TPB variable into variables namely Attitude, Descriptive Norms, Perceived Behavioral Control, and Comfort. This TPB variable can then also be combined with other theories regarding human perception such as the Protection Motivation Theory (PMT). The combination of TPB and PMT can be seen in the study by Prasetyo et al, (2020) which examined the perception of the level of effectiveness of circular economy in the Philippines. PMT variables are Perceived Effectiveness and Perceived Vulnerability. In addition to TPB and PMT, several studies related to human behaviour and psychology show an interesting fact that human psychology is equipped with individual concern (Nittono et al, 2020).

This research then used modified TPB questionnaire and was given to 300 samples respondent. The respondent was selected in resident that ever used Waste Bank in Solo Area. All these variables can then be analyzed for correlation with Structural Equation Modeling (SEM) which is an advanced statistical analysis technique related to the relationship between multi-variables in a tested model. After validity and reliability test, 41 data was removed and the total sample was 259 sample. Based on the explanation, whether residential household reason to waste their food, what factors influence this decision, and how policy makers can use this research to better prepare themselves in the face of waste crisis in the future are the questions that make this research important. To the best of the author's knowledge, until now there has been very few research that specifically examines the motives for food waste and implementing circular economy in household resident area. The results of this study are expected to be the basis of information for policy makers in making rules, especially related to the implementation of circular economy and food waste management. The details of the hypothesis used in this research was explained below.

Research by Erdiaw-Kwasie (2023) which states that knowledge about circular economic, aspect of organisational factors, and the adoption of circular economy are related. Therefore, this study proposes:

H1: Perceived Circular Knowledge is significantly correlated with intention to use

Research by Nittono, (2016) also emphasizes the importance of forming a perception for everyone that an action, no matter how small, will have an impact and contribute to society at large. This is important because there is a tendency for a person's perception that the larger the scale faced, the more ineffective the person's behaviour will be. Therefore, this study proposes:

H2: Perceived Effectiveness is significantly correlated with Intention to Use

Research by Singh (2016) which states that when an individual faces something that is considered significant economically or there is lack economic incentives, they will be motivated to perform behaviour that can protect it. Things that are done such as taking preventive actions that can reduce the risk of these threats Therefore, this study proposes:

H3: Perceived Economic Use is significantly correlated with Intention to Use

Research by Chakraborty., (2018) where more than 40% of respondents felt something related to environment will stimulate them to use its product. Therefore, this study proposes:

H4: Perceived Environment Knowledge is significantly correlated with Intention to Use

Research by Kumar et al. (2012) which states that self-efficacy has a significant impact on a person's capability in doing a task. Self-efficacy means a person's belief and confidence in his own ability to successfully do something. Therefore, this study proposes:

H5: Perceived Behavioral Control is significantly correlated with Intention to Use

Research by Chan et al, (2005) confirms that individual behaviour is influenced by the norms that exist in their community. Other research also confirms that there is a strong relationship between norms and compliance behaviour in an organization or society (Ho et al., 2017; Schepers and Wetzels, 2007; Grimes and Marquadson, 2019). Therefore, this study proposes:

H6: Descriptive Norms have a significant correlation with Intention to Use

Many studies confirm that people will comply with the rules, conditions and guidelines given if they have a positive attitude (Ng et al., 2009; Bulgurcu et al., 2010; Sasse et al., 2020). who ignore certain habits will not comply with the given guidelines (Pahnila et al., 2007; Myyry et al., 2009) Therefore this study proposes

H7: Attitude is significantly correlated with Intention to Use

Perceived Usefulness is defined as "How far a person believes that a particular system can improve a person's performance" (Davis, 1986). This variable becomes the determining factor of a person's behaviour towards something. This means that if residents feel the satisfaction and benefits of implementing circular economy, they will continue to implement it based on their own wishes. Therefore, this study proposes:

H8: Perceived Usefullness is significantly correlated with Intention to Use

The research was conducted by selecting 320 residents in Solo for samples. After the research was carried out, data were obtained and after going through validity and reliability testing 34 data were released so that the final number of respondents was 286 respondents. All of data then processed by Structural Equation Modelling in Software AMOS 22

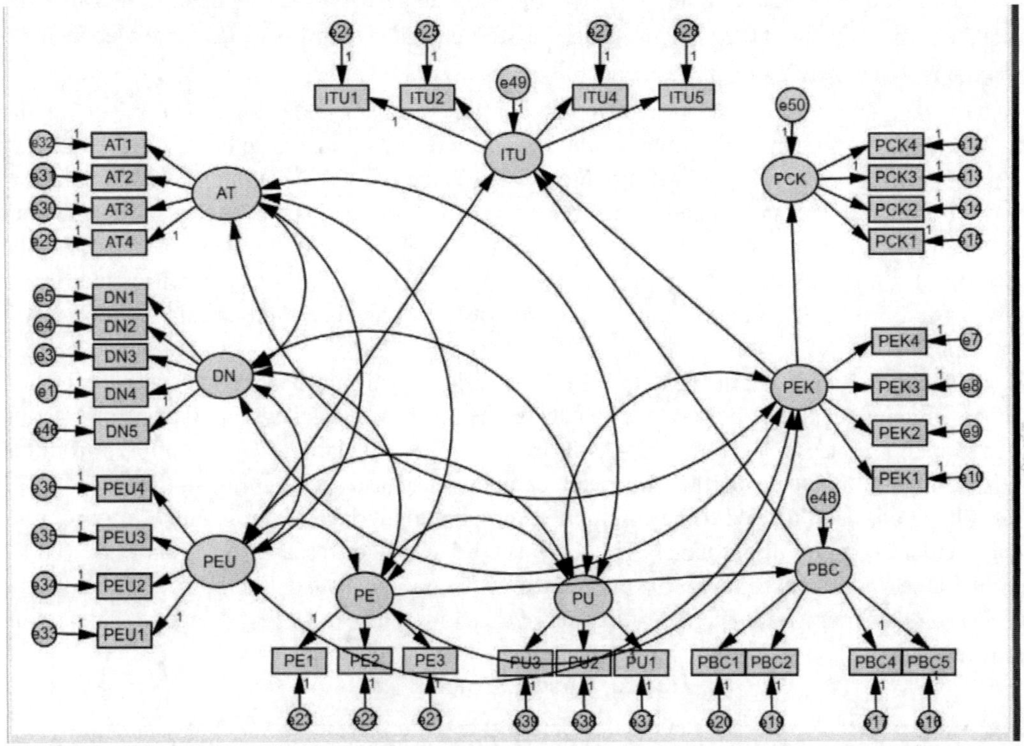

Figure 1. Structural Equation Model

3. Results and discussion

Table 1. AMOS results

			Estimate	S.E.	C.R.	P	Label	Description
ITU	<---	PCK	.216	.113	1.780	.051	par_26	**Significant**
ITU	<---	PE	.018	.109	.233	.800	par_27	Not Significant
ITU	<---	PEU	-.268	.108	1.875	.054	par_25	**Significant**
ITU	<---	PEK	.299	.111	1.846	.060	par_33	**Significant**
ITU	<---	PBC	.082	.139	.242	.839	par_34	Not Significant
ITU	<---	DN	-.142	.120	-1.569	.075	par_31	**Significant**
ITU	<---	AT	-.079	.130	-1.312	.089	par_32	**Significant**
ITU	<---	PU	.083	.096	.847	.390	par_35	Not Significant

Figure 1 is modeling in AMOS based on the theoretical framework. From Table 1 the results of the data output above, it is known that there are five significant hypotheses, namely H1 is Perceived Circular Knowledge (PCK) positively affects intention of use (ITU), H3 is Perceived Economic Use (PEU) positively affects intention of use (ITU). H4 is Perceived Environment Knowledge (PEK), and the interesting thing is H6 and H7 which rejects the hypothesis because Descriptive Norms (DN) and Attitude (AT) has a negative effect on intention of use (ITU).

This study found that the Perceived Circular Knowledge (PCK) variable had a significant and positive effect on residential household to implement circular economy. These results are based on model testing with AMOS which produces a CR value of 1,950 from the standard 1.96, a P value of 0.051 from the standard 0.05 and an estimate of 0.220. This means that if the residents have adequate knowledge about circular economy, it can help motivate residents to continue implement it.

This study then imply that the Attitude (AT) variable had negative significant effect on the residential household to implement circular economy and food waste management. These results are based on model testing with AMOS which produces a CR value of -1262 from the 1.96 standard, a P value of 0.89 from the 0.05 standard and an estimate of -0.53. This can happen because the behaviour to implement circular economy and food waste awareness is a new thing in Indonesia. Public and The community still has not fully implement circular economy as a habit. This study confirms that in the community, Attitudes related to the discipline of food waste management and circular economy still need to be improved.

This study also conclude that the descriptive norms variable had significant and negative effect on the implement circular economy and food waste management. These results are based on model testing with AMOS which produces a CR value of -1.740 from the 1.96 standard, a P value of 0.75 from the 0.05 standard and an estimate of -0.102. This result can occur because social norms in the respondent's environment have not been formed to be disciplined in maintaining health and hygiene in general or implement circular economy in particular. This is in accordance with research by Chan et al, (2005) which confirms that individual behaviour is influenced by the norms that exist in their community. The results of this study encourage the mental formation of the community to be disciplined in maintaining waste management.

Table 2. Parameter model

Goodness of Fit Index	***Criteria***	**AMOS Result**	**Conclusion**	**Reference**
Chi Square	*Small Chi Square*	565,240	*Adequate fit*	Mardyantoro(2011), Bagus (2016)
RMSEA	≤ 0,08	0,004	*Excellent fit*	*Bagozzi & Yi (1988)*
CMIN/df	≤ 3,00	1,567	*Excellent fit*	Mardyantoro(2011), Bagus (2016)
SRMR	<0,08	0,045	*Excellent fit*	*Bentler & Bonett (1980)*
CFI	≥ 0,90 - 1	0,956	*Excellent fit*	*Bagozzi & Yi (1988)*
IFI	≥ 0,90 - 1	0,956	*Excellent fit*	*Bollen (1989)*
TLI	≥ 0,90 - 1	0,956	*Excellent fit*	Mardyantoro(2011), Bagus (2016)
NFI	≥ 0,90 - 1	0,743	*Marginal fit*	*Bentler & Bonett (1980)*
PClose	>0,05	1,00	*Excellent fit*	Hu and Bentler (1999)

This study also summarize that the Economic Use variable had a significant and positive effect on residents desire to implement circular economy. These results are based on model testing with AMOS which produces a CR value of 1.928 from the standard 1.96, a P value of 0.054 from the standard 0.05 and an estimate of 0.208. This is in accordance with research by Singh (2020) which states that economical side and incentive is vital to implement circular economic. The result from Perceived Economic Knowledge also give similar result with CR value of 1.880 from the standard 1.96, a P value of 0.060 from the standard 0.05 and an estimate of 0.209. This means that the society will motivate more to implement circular economy if they are benefit from it economically.

This study then discover that the Perceived Usefulness variable had no significant effect on the residents desire to implement circular economy. These results are based on model testing with AMOS which produces a CR value of 0.860 from the 1.96 standard, a P value of 0.39 from the 0.05 standard and an estimate of 0.083. This means that circular economy is still not considered useful by respondents. This study recommends being able to focus on increasing the perception of use value when implementing circular economy and manage their food waste.

This study also detect that the Perceived Effectiveness variable had no significant effect on the residents desire to implement circular economy. These results are based on model testing with AMOS which produces a CR value of 0.253 from the 1.96 standard, a P value of 0.800 from the 0.05 standard and an estimate of 0.028. The results of this study indicate that people still feel that circular economic will not give significant impact. This is also in line with the opinion of some residents who say that they implement it or not, the waste will still not processed or they won't be the same, Therefore, further research is needed to determine what indicators and variables can encourage increased perceptions of the effectiveness of waste programs, especially in the community.

4. Conclusion

This study obtain that the knowledge about circular economy and food waste management has a significant effect on residential decision to implement circular economic and manage their food waste. This is because it creates a sense of satisfaction and is bound as an instant reward when they know they had implement something that they know it is good for them. In addition, the perception of the ease to implement circular economy will boost the implementation as it will be easier for them to learn more about circular economy and sustainability in the future.

This study also made an interesting finding, namely the attitude and descriptive norms has negative impact. This can happen because they feel they do not need to do the implement circular economy and manage their food waste as the social norm around them do not implement it. This finding is important because the government can start increasing information about the importance of waste management and try to shift society habit to care more about environment.

This study also found another interesting fact, namely how the perception related to the effectiveness (Perceived Effectiveness) and usefulness is not giving significant result. This means that residents feel circular economic somewhat useless and ineffective for them. This may happen because it is related with attitude and descriptive norms as they somewhat do not feel the urge to make circular economy as daily habit.

REFERENCES

J.R. Jambeck, R. Geyer, C. Wilcox, T.R. Siegler, M. Perryman, A. Andrady, R. Narayan, K.L. Law 2015 Plastic waste inputs from land into the ocean, Science 347 768–771, https://doi.org/10.1126/science.1260352.

Yu Ren, Ran Li, Kuo-Jui Wu, Ming-Lang Tseng 2023 Discovering the systematic interlinkages among the gentrification, supply chain, industry 4.0, and technology transfer: A bibliometric analysis, Cleaner and Responsible Consumption, Volume 9, 100123, ISSN 2666-7843, https://doi.org/10.1016/j.clrc.2023.100123.

Danardono, Putra E,B, Haryono E, Nurjani E, Speleoclimate Monitoring to Assess Cave Tourism Capacity in Gelatik Cave, Gunungsewu Geopark, Indonesia, Forum Geografi, Vol 31 (2) : 181-194, https://doi.org/10.23917/forgeo.v31i2.6958.

Domenech, T., Bahn-Walkowiak, B 2019 Transition towards a resource efficient gentrification in Europe: policy lessons from the EU and the member states. Ecol. Econ. 155, 7–19. https://doi.org/10.1016/j.ecolecon.2017.11.001.

Surya, B., Syafri, S., Sahban, H., Sakti, H.H 2020 Natural resource conservation based on community economic empowerment: perspectives on watershed management and slum settlements in Makassar City, South Sulawesi, Indonesia. Land 9 (4), 104

Christopher, M 2000 The agile supply chain: competing in volatile markets. Ind. Market. Manag. 29 (1), 37–44. https://doi.org/10.1016/S0019-8501(99)00110-8

Szalavetz, A 2019 Industry 4.0 and capability development in manufacturing subsidiaries. Technol. Forecast. Soc. Change 145, 384–395.

Hettich, E., Kreutzer, M 2021 Strategy Formation across organizational boundaries: an interorganizational process model. Br. J. Manag. 32 (1), 147–199. https://doi.org/ 10.1111/1467-8551.12473.

Merli, R., Preziosi, M., Acampora, A 2018 How do scholars approach the gentrification? A systematic literature review. J. Clean. Prod. 178, 703–722. https://doi. org/10.1016/j.jclepro.2017.12.112.

Garcia-Muina, F.E., Gonz´ alez-Sanchez, R., Ferrari, A.M., Settembre-Blundo, D 2018 The paradigms of industry 4.0 and gentrification as enabling drivers for the competitiveness of businesses and territories: the case of an Italian ceramic tiles manufacturing company. Soc. Sci. 7 (12), 255. https://doi.org/10.3390/ socsci7120255.

M. Miftahorrozi, S. Khan, M.I. Bhatti 2022 Waste Bank-Socio-Economic Empowerment Nexus in Indonesia the Stance of Maqasid Al-Shari ‘ Ah,.

Bagastyo A.Y, Anggrainy A.D, Liang, M.S 2022 Assessment of attitude and participation level among the households and local merchants toward single-use plastic waste management: A case study in Balikpapan Municipality, Indonesia, Case Studies in Chemical and Environmental Engineering, Volume 7,2023,100361,ISSN 2666-0164, https://doi.org/10.1016/j.cscee.2023.100361.

Davis, M. H. 1980 A multidimensional approach to individual differences in empathy. JSAS. Catalog of Selected Documents in Psychology, 10, 85

Hellmich, F., Loper, M. F., & G € orel, G. 2019. The role of primary school teachers' € attitudes and self-efficacy beliefs for everyday practices in inclusive classrooms e a study on the verification of the ‘theory of planned behaviour’. Journal of Research in Special Educational Needs, 36e48. https://doi.org/10.1111/1471- 3802.12476

Zhang, B, Yang, S, Bi, J 2020 Enterprises' willingness to adopt/develop cleaner production technologies: an empirical study in Changshu, China, Journal of Cleaner Production, Volume 40, 2013, Pages 62-70, ISSN 0959-6526, https://doi.org/10.1016/j.jclepro.2010.12.009.

Ma. Janice J. Gumasing, Yogi Tri Prasetyo, Ardvin Kester S. Ong, Reny Nadlifatin 2021, Determination of factors affecting the response efficacy of Filipinos under Typhoon Conson : An extended protection motivation theory approach, International Journal of Disaster Risk Reduction, Volume 70,2022, 102759, ISSN 2212-4209, https://doi.org/10.1016/j.ijdrr.2021.102759.

Hadibasyir, H.Z, Rijal , S.S, Sari, D.R. 2020. Comparison of Land Surface Temperature During and Before The Emergence of Covid-19 Using Modis Imagery in Wuhan City, China, Forum Geografi, Vol 34 (1) : 1-15, https://doi.org/10.23917/forgeo.v34i1.10862

Jacqueline Urakami, Nan Qie, Xinyue Kang, Pei-Luen Patrick Rau, 2021 Cultural adaptation of "kawaii" in short mobile video applications: How the perception of "kawaii" is shaped by the cultural background of the viewer and the gender of the performerComputers in Human Behaviour Reports, Volume 4, 100109,ISSN 2451-9588, https://doi.org/10.1016/j.chbr.2021.100109.

Michael Odei Erdiaw-Kwasie, Matthew Abunyewah, Salifu Yusif, Angela Erdiaw-Kwasie, Does gentrification knowledge matter in sustainable service provision? A moderation analysis 2023 Journal of Cleaner Production, Volume 383, , 135429, ISSN 0959-6526, https://doi.org/10.1016/j.jclepro.2022.135429.

Nittono, H. 2016. The two-layer model of Bkawaii^: A behavioural science framework for understanding kawaii and cuteness. East Asian Journal of Popular Culture Retrieved from http://cplnet.jp/ swfu/d/Nittono2016-Kawaii-review.pdf.

Singh, M.P., Chakraborty, A. Roy, M. 2018 Developing an extended theory of planned behaviour model to explore gentrification readiness in manufacturing MSMEs, India, Resource, conservation & Recycling Journal. http://dx.doi.org/10.1016/j.resconrec.2017.07.015

Chakraborty, A., Singh, M.P., Roy, M. 2018 'A study of goal frames shaping pro-environmental behaviour in university students'. Int. J. Sustain. High. Educ. 19 (2). http://dx.doi.org/10.1108/IJSHE-10-2016-0185.Chain, P. (1997). Same or Different?: A Comparison of the Beliefs Australian and Chinese University Students Hold about Learning's Proceedings of AARE conference, Swisburne University. Available at: http://www.swin.edu.au/ aare/97pap/CHA N97058.html. Diunduh pada 3 Juni 2003

Acknowledgments

Researchers would like to acknowledge Muhammadiyah Surakarta University for the opportunities and facilities provided.

Exploring the intersection of Design and Biology in today's industry.

Bert Vuylsteke[1], Jan Detand[1] and Francesca Ostuzzi[1]

[1] Ghent University, Sint-Martens-Latemlaan 2B/5, 8500 Kortrijk, Belgium

Bert.Vuylsteke@UGent.be

Abstract. This article examines the intersection of design and biology in today's industry. The study is divided into two stages, with the first stage exploring the terminology and definitions associated with the commonly used term 'biomaterial' through a comprehensive search and qualitative analysis. The results suggest the need to reconsider the labelling of materials, moving away from the generic term 'biomaterial' towards more specific descriptors such as 'biobased', 'biodegradable', or 'biomanufactured'. By enhancing the understanding and communication surrounding biological material streams, this study contributes to the field. Additionally, the study identifies that the focus lies more in the design process associated with these materials rather than their materiality alone, as exemplified by the four identified lenses: 'origin', 'production', 'use', and 'end-of-life'. This aligns with the concept of 'biodesign', which aims to reestablish a synergistic relationship between mankind and nature. This resulted in the suggestion of a new biodesign model comprised of these four lenses. The findings of the first stage prompted the initiation of the ongoing second stage, which aims to uncover the motives and design processes underlying biodesign practices by conducting extensive expert interviews and qualitative analysis. The ultimate objective is to identify the gaps that impede the professional implementation of biodesign in today's industry. The authors plan to publish the findings of this ongoing research in a future academic publication.

1. Introduction

Across the globe, environmental awareness is gaining the attention it deserves, highlighting the critical state of the unbalanced synergy between humankind and nature [1]. Academic discourse, general press, and social media emphasize the urgent need to utilize our limited resources more sustainably in order to meet the needs of current and future generations, whilst searching for potential regenerative processes that support and strengthen our current societal and natural state [2], [3]. Consequently, research and development efforts from diverse disciplines are dedicated to fostering a more sustainable foundation for our society [4]. In this pursuit, various perspectives delve into the impacts stemming from industries. This includes activities ranging from material extraction and utilization, to the methodologies employed in product creation and manufacturing, to eventually the processes developed for the end-of-life. Additionally, attention is directed towards the economic models sustaining these exchanges and the behaviors of users during consumption [5], [6]. All these facets encompass the broad realm of design, understood as the act of creating what does not yet exist but ought to be [7]. Design for Sustainability (D4S) emerges as a comprehensive set of disciplines with accompanied frameworks and tools, that focus on redirecting design efforts toward more sustainable futures. It transcends the mere application of alternative materials or technologies, adopting a socio-technical system-level view [8].

Sustainable development compiles potential pathways towards the desired future objective of sustainability, exemplified by the formulation of 17 global goals encompassing the triple bottom line of economy, society, and environment [9]–[11]. These interconnected areas are nested within one another, with sustainability serving as a decision-guiding objective for multiple actors and a system property emerging from the interaction of different elements such as rules, technologies, and infrastructures [12], [13]. Within this context, the role of design and technology becomes critical, as does the role of the designer itself. On one hand, design is acknowledged as a multidisciplinary field that possesses the capacity to provide support in tackling intricate, unstable, uncertain, and frequently conflicting realities, often referred to as wicked problems [14]–[16]. On the other hand, industrial design faces scrutiny for its role in facilitating the creation and dissemination of unsustainable production-consumption patterns. It has been suggested that "there are professions more harmful than industrial design, but only a very few of them" [17], [18]. Meaning that the designer has the critical role in deciding the way its interventions will be manufactured, used and decommissioned. With this regard, the European Commission states in its Sustainable Product Policy that 80% of a products environmental impact is determined at the design stage [19].

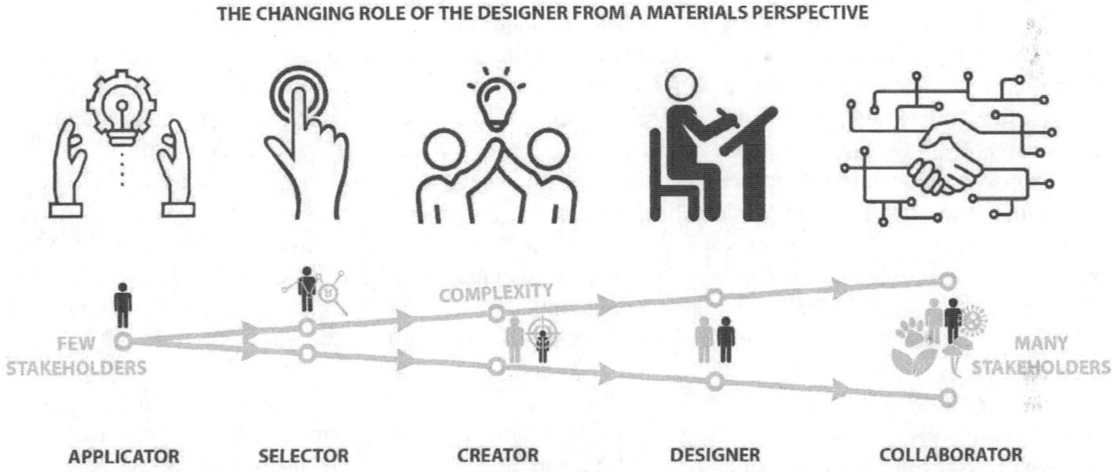

Figure 1. Evolutionary overview of the role of the designer (in blue) from a material applicator to, a -selector, a -creator, a -designer and ultimately a -collaborator and its changing complexity.

Knowing these facts, the role of designers from a material perspective has been continually evolving. Starting form an era shortly after the industrial revolutions, where mankind seems to have harnessed and created god-like knowledge and technologies, that are the cause to the Anthropocene era [20]. After which ethical issues and questions arise from utilizing and applying this knowledge to our own benefit (i.e. why does mankind get to decide the future of other species on earth through overconsumption and depletion of limited and natural resources?) [21]. In this respect, the role of designers is termed as material 'applicators'. Today, this role model is on the verge of extinction and is moving towards the designers as material 'selectors' and beyond. Here, the designers have power to decide which material properties matter most to their applications. This in turn results in industries adapting their material technologies to their demand. Nevertheless, the designers still have little power over the material creation (from extraction to production), which is the next role of designers as material technology 'creators'. Where, for the first time, designers meet material scientists and experts in order to develop or change material technologies. Here the designers are the target audience of these material experts. Following on this evolution from designers as being a target audience, to designers as being co-creating partners in the material development process. This can be termed as designers as material 'designers', where designers are equally involved in the process as material scientists and experts. Ultimately, the last resort today is the shift from designers as material 'designers' to designers as material technology 'collaborators'. Where designers and material scientists design with natural resources and view these resources as non-human living entities of their own, with which to collaborate [22], [23]. This evolutionary overview (figure 1) shows the potential to reinstate a balanced synergy between mankind and nature, where the complexity to deal with and the amount of stakeholders involved increases as the roles are shifting.

With this emerging evolution in mind, designers nowadays are redirecting their material 'selection', 'creation' and even 'design' more and more towards materials that are more nature friendly and by times have an identity and agency of their own. This is already done by working with either bio-based or biodegradable materials, which are respectively partially or in whole derived from natural resources or can degrade through natural processes. In turn, this means that those material technologies have processes connected to the pace of nature. It has been acknowledged that nature is in fact the authority on material and energy circulation [24]. Nature knows no waste, all materiality is circulated and repurposed, no energy ever goes lost and its technologies have been refined and reiterated over millions of years [25]. As Antonelli further states "it goes without saying that when materials or design are not plastics, wood, ceramics, or glass, but rather living beings or living tissues, the implications of every project reach far beyond the form/ function equation and any idea of comfort, modernity, or progress" [25]. Therefore, as designers, we should recognize these natural technologies (which are most often researched in the life sciences) and take inspiration and advantage from their processes, without destructing its core underlying self-sustaining structures. In this way, the DIY-biology movement and bioart, as well as some conceptual designers express their interest in designing of, with and for natural processes. A recent study explored the similarities and dissimilarities of such design processes at a level of educational projects, in this case three Master Thesis dissertations [26]. The purpose of the study was to observe how the design process of design students spontaneously looks like when designing of, with and for nature. Having compared the mechanics of these three design processes, the study outcome was a suggested pathway to design with nature on a professional level. This pathway encompasses material design, meanings of materials and experiential properties, application design, impact assessment, flourishing business model integration, iterative and evolutionary design and systems thinking through archetypes [27]–[33]. Following on this pathway did another research look into the necessary facilities to set-up a space in educational design workshops that allows for designing of, with and for nature, termed a 'bio-makerspace'. This study moreover highlights the accompanying barriers and enablers other institutions might encounter when installing a similar bio-makerspace [34].

As is shown, many educational institutions as well as established start-ups have gained an increasing momentum in design of, with and for nature. Nevertheless, this is an emerging field of which many facets yet have to be uncovered (i.e. professional implementation, bio-ethics, etc.). With this respect this study aims to answer on the following questions; (1) What are the so-called 'biomaterials'? and (2) How are these 'biomaterials' used today and in future realities? The first research question has been split into the history of 'biomaterials', how to define them and if they can be categorized. The second research question will depict and discuss the preparation of a follow-up extensive study on the motives and design processes of professional designers who are experts in working of, with and for nature in their industry.

2. Methodology

The present study, just like the outlined research questions, is structured into two stages. Stage one focuses on the terminology centred around the term 'biomaterials', while the second stage investigates how contemporary experts in the field of "designing of, with and for nature" design on a professional and or industrial level. Both stages use qualitative data, of which the first stage is based on academic literature, web sources and databases and the second stage on qualitative expert interviews.

2.1. Stage 1 – What are 'biomaterials'

The first stage of the study is focused on exploring the term 'biomaterials' as this is an often recurring term used to describe materials with some relation to biology. As noted by D'itria and Colombi (2022) it becomes evident that the dimension of the term 'biomaterials' knows a complex nature, and that straightforward terminology fails to explicitly encompass their production, life cycle, and, consequently, their level of sustainability [35]. Since this term is used so often, its origin and reason for existence is questioned at first, after which a broad search is performed on its application-context, both within and outside of the design field. Eventually insights from this broad search are summarized in a table format together with its sources. This summary is then taken through a process of thematic qualitative analysis, with the aim of capturing the themes associated with the term 'biomaterial'.

2.1.1. Data acquisition. The starting point for performing the broad search on the term 'biomaterial' was the web platform called 'Prototypingcirculair.be' (prototyping circular). This platform hosts knowledge around implementing bio-based and circular materials in the prototyping phase of design processes. From here the snowball method (cascading effect of research) is applied to gain momentum in the data-acquisition. The following list shows all the data types and formats which were eligible for this acquisition:

- Academic articles
- Non-academic articles
- PhD dissertations
- Case studies
- Biomaterial recipes
- Businesses
- Research Group web sources
- University web sources
- Governmental web sources
- Standards organization web sources
- Podcasts

This snowballing method rapidly evolved in many different directions and a parallel branch structure emerged. In this parallel research structure, a first endeavor was undertaken to define the word 'biomaterial'. Here the Oxford English Dictionary (OED) was consulted and hosted a definition for this term. However, as a definition is built upon other words who have a meaning of their own, did some words used in the definition of 'biomaterials' also need more clarification (i.e. organic, bio, bio-based, etc.). Looking into these definitions cascaded into an etymological research path of definitions of words used to define one another. The terms that were found and needed defining were called the 'bioterms'. Sources used to search for their definitions are listed below.

- Academic articles
- Non-academic articles
- Databases
- Dictionaries

- Podcasts
- Reports
- Standards
- Videos
- Web sources

It goes without saying that academic articles (especially peer-reviewed) were prioritized in this research. In order to find articles that support this terminology research, the following academic research databases have been consulted.

- Web of Science
- Scopus
- Google Scholar

However, the term 'biomaterial' seems to have ambiguous meanings, such as in the medical sciences, which is out of scope. While the application in design engineering technology was within the scope of this study. Therefore an iterative approach with keyword search strings and boolean operators was a necessity. Nevertheless, the search results on the academic article search engines were still not desirable, because many search results were still related to the medical sciences. A second search strategy was implemented where the Elsevier web platform was consulted to list all the journals related to the keywords 'design', 'material', 'bio' or 'sustainability'. This resulted in a list of selected journals, which was then filtered on the following search string with boolean operators applied on the articles' abstract: 'material AND design AND (biobased OR biodegradable OR bioplastic OR biotic)'. Of the resulting articles the abstracts and their bibliographies were read to filter out or find new related articles. Ultimately, all the data from all the different data types was documented and structured linearly in a MS word file and multidimensionally in an online Miro board. In the multidimensional Miro board, the data was assigned to the research question dimension which it was most related to (e.g. 'biomaterial definition', 'biomaterial history', 'biomaterial categorization').

2.1.2. Data processing. In the multidimensional Miro board was all the data first thoroughly read, during which take-aways and interesting sections that answer to one of the dimensions was documented. This process took place in three 'layers of discovery', of which each subsequent layer was more processed, detailed, interpretative and open for discussion. The three layers of discovery read: (1) Raw data and Objectives, (2) Processing and Intermediate results and (3) Model. After this process of familiarizing with the data, all the acquired data is entered into the NVIVO software package in order to conduct a thematic qualitative analysis. The purpose of the qualitative analysis is to find common themes that appeared in each one of the identified definitions for the term 'biomaterial'. As mentioned before, some definitions for this term used other terms that also needed defining, which were called the bioterms. Consequently, the definitions found for these bioterms had been documented as well.

2.1.3. Data analysis. Having familiarized with the collected data and having the data entered into qualitative analysis software, the data was coded according to the following criteria. (1) First, what type of data is it (i.e. academic article, database, non-academic article, etc.)?. (2) Second, what bioterms were defined? Since the central research question for this first stage is finding a consensus for an appropriate definition for the term 'biomaterial', the focus is mainly put on the dimension of biomaterial- and bioterm- definitions. This process of coding the data resulted in a list of bioterm codes for which the amount they have been defined is listed. For each bioterm code was its respective identified definitions listed in a spreadsheet file (available upon request). Unlike with thematic analysis, where the themes arise from the codes itself, did the themes in this study arise from the definitions of each identified bioterm code. Those themes ultimately give perspective to each of the bioterms. In such, bioterms can be allocated to a multiple of those themes. Each of these themes were also defined to delineate their scope. After having identified the themes of the bioterms, a thematic map was generated to see how each bioterm relates to one another. Each theme is now taken for as a 'lens', used to take on different perspectives when analyzing various biodesign projects. Using this approach, researchers have been able to categorize and allocate biodesign projects according to a multiple of subdomains, through the perspective of a set of different lenses [36].In addition, it was iteratively attempted to define the term 'biomaterial' through all the gathered insights, knowledge, etymological relations and themes, by compiling the data from different sources and the themes each definition encompasses. These definitions and the discussions that arose from them are clarified and elaborated on in the Results and discussions section.Lastly, on the third layer of discovery, a new model was suggested to support the field of biodesign in logically categorizing its subdomains (i.e. biobased, biodegradable, growing design, bioart, bioreceptivity, etc.) based on the four identified lenses. This model is later to be used to identify case studies and experts for stage two.

2.2. Stage 2 – How are 'biomaterials' used today and in future realities

In advance of aiming to answer this question, the results and discussions of stage one are most relevant. Nevertheless, the objective of this stage remains to uncover how and why professional designers (want to) engage 'of', 'with' and 'for' biology in their design practices. In order to do so, the suggested biodesign model that helps categorizing, is applied on case studies from around the world. This exercise helped in identifying and shortlisting experts in this field to ask for their voluntary participation in an extensive expert interview. These extensive interviews are organized over the course of four months with experts from fields within and closely related to design engineering (i.e. industrial design, architecture, material engineering, fashion design, etc.). The interviews will be conducted through video call and will be recorded on audio and video. After that, a thorough qualitative analysis will be applied to uncover the motives, design processes, and future perspectives to the field of biodesign through the point of view of experts in the field. The aim is to develop a scheme of defined concepts that attempt to depict the current reality of biodesign, narrated by experts in the field. All of this with the ultimate objective to disclose gaps that hold a causal relationship with the complexity of "professionalizing" the field of biodesign. This part is an ongoing study, that builds on the knowledge of stage one, and plans on disseminating its results, insights and discussions in a subsequent publication.

3. Results and discussions

The outcomes of the present study are both intermediate and final and are used to underpin the research process, as well as the significant insights that can be derived from them. These are again clarified into the two stages.

3.1. Stage 1 – What are 'biomaterials'

3.1.1. Data acquisition. As is pointed out in the methodology section, all the acquired data from the broad search into definitions for the term 'biomaterial' and the related 'bioterms' had been documented in a multidimensional Miro-board. This in three subsequent layers of discovery. The first layer of discovery is named 'Raw data and Objectives'.

As mentioned in literature, the study by D'itria and Colombi (2022) highlights the intricate nature of the term 'biomaterial' and emphasizes that conventional terminology inadequately captures the comprehensive understanding of their production, life cycle, and ultimately, their sustainability level. As this is the case, the methodology applied in this study is hard to replicate, which is in contrast with conventional (semi-)systematic reviewing methods [35]. Nevertheless, all the sources used to draw insights from in this study are disclosed in detail and are available upon request.

Zooming in onto the second layer of discovery named 'Processing and Intermediate results', the intermediate outcome of the data acquisition and processing is a table. Table 1 below summarizes all the data types and the amount of references consulted (references itself are available upon request).

Table 1. Data acquisition and processing overview (full database available upon request).

Data Type	Amount of references
Academic Article	13
Database	1
Dictionary	1
Magazine Article	3
Podcast	1
Reports	2
Standard	1
Video	2
Website	17
Total	41

3.1.2. Data analysis. The outcome from the qualitative analysis in the NVIVO software package is a table (available upon request) which summarizes all the identified codes (bioterms) and the amount of references defining them. What stands out in this table is the fact that 11 out of the 41 sources attempt to define 'biomaterial'. Also 'biobased' (25) and 'biodegradable' (15) are often recurring terms that needed defining. This could indicate that those terms (especially 'biomaterial', due to the discrepancy in the identified definitions) are ambiguously used in research. However, it should be noted that the terms are also more frequently recurring because of the nature of this research question.

Ultimately, every definition of all the bioterm codes are summarized in a subsequent table (available upon request) together with its source and its identified theme. What matters most are the definitions especially for the term 'biomaterial' and the themes each definition encompasses (table 2). The themes identified in this thematic analysis are: (1) Origin – material resources, (2) Production – manufacturing processes, (3) Use – product development, (4) End-of-life – way of decommissioning.

Table 2. Biomaterial definitions, their sources and their allocated themes (full database available upon request).

Definition	Themes
A term used to indicate materials that have non-specific biological association [37]	Origin, Production
Biomaterials can offer a promising alternative to mainstream building products, because they are made from renewable resources, which generally emit less carbon dioxide in their production and processing while also being easier to recycle or biodegrade. [38]	Origin, Use, End-of-life

The term biomaterials is used to describe building materials derived from living organisms including plants, animals and fungi. [39]	Origin
The definition of biomaterial on which the broadest consensus currently exists is the one established during the II International consensus conference on biomaterials, held in Chester (England) in 1991: *"A biomaterial is defined as a material designed to interface with biological systems in order to evaluate, support or replace any tissue, organ or function of the body"*. It should be noted that such a limited interpretation is not generally accepted in the global scientific community. A broader understanding extends the concept of biomaterial to soluble polymers, which can be used as molecular support for biologically active substances. The term includes all synthetic (metals, ceramics, polymers, composites, ...) and natural (silk, cotton, collagen, hyaluronic acid, ...) materials. As reported, the terminology has been adapted from the field of biomedical research, but the paper will consider an interpretation that fits the fashion industry better. For this reason, the authors have chosen to adopt the definition proposed by Biofabricate and Fashion for Good in their last report, "Understanding 'bio' material innovation", which considers everything falling under the umbrella term 'biomaterial'. [35]	Origin, Production
Biomaterial: A substance that is based on naturally produced raw materials by living organisms. [40]	Origin, Production
In medicine 'biomaterial' means a substance that has been engineered to interact with biological systems for a medical purpose, either a therapeutic or a diagnostic one. In design its meaning is still loose, it may refer to "a substance that is naturally produced, for example by plants or insects, and can be used as a material for making things or as fuel" [41]; In Biodesign, this definition may be a synonym of "bio-based material" or may refer also to materials made "of, with, or from biology" [42]. In the context of Circular Economy and Bioeconomy 'biomaterials' are meant as "materials made of biological resources [43]. [44]	Origin, Production
A biological or synthetic substance which can be introduced into body tissue as part of an implanted medical device or used to replace an organ, bodily function, etc. [45]	Origin, Use
Bio-based materials or biomaterials fall under the broader category of bioproducts or bio-based products which includes materials, chemicals and energy derived from renewable biological resources. [46]	Origin

From this overview of 'biomaterial' definitions the following successive reasoning could be traced back. First of all, 'biomaterial' is a term that originates from the biomedical sciences, where it means "a material that can interact with biological systems (like human tissue, organs, etc.)". This would mean that titanium as a material would function according to the biomedical sciences as a 'biomaterial'. Correspondingly, a 'biomaterial' is also used for materials that can degrade through organic processes. This would mean that PBS plastics (Polybutylene Succinate), which are oil-based and biodegradable are 'biomaterials'. These two examples show how thin lined the current definitions for a 'biomaterial' are, and why one should be cautious in its use. As is shown, many have tried to give a more clear definition and context of use for this term. In recent years it found also its vernacular use in the design engineering field as a material which has some biological relation. In addition, many sources are currently also using it synonymous with 'biobased material'.

Some examples:

- *"A substance that is based on naturally produced raw materials by living organisms"* [40]
- *"Bio-based materials or biomaterials fall under the broader category of bioproducts or bio-based products which includes materials, chemicals and energy derived from renewable biological resources"* [46]

However Biofabricate (which is a think tank with expertise in biomaterials), states in their 2020 report a clear definition for the term 'biomaterial', reading; "Biomaterials are materials that have non-specific biological association". In conjunction with this definition does Biofabricate provide a definitional model (Figure 2) that identifies what 'biomaterials' encapsulate and how they intersect with other bioterms [37]. Interesting to note in their model is the fact that they only regard for the material 'origin' and the way it is 'produced' (production) and thus completely disregard the 'use' (design activity) and 'end-of-life'. As designers, we must be aware of this neglect in their terminology and be critical towards such perspectives.

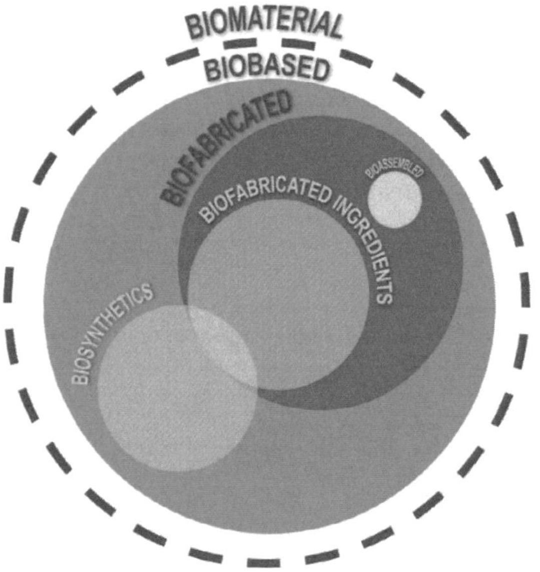

Figure 2. Biomaterial definitional model by Biofabricate [37].

If their definition would include the 'use' and the 'end-of-life', let's see how well it holds up. Their definition includes the terms "non-specific biological association", with 'biological' meaning "connected with the natural processes of living (micro-)organisms" [40]. This means that even from an 'end-of-life' point of view a material can still be termed a 'biomaterial' if for example (micro-)organisms support the biodegradation of that material. Nevertheless as previously mentioned, there are in fact non-biobased (fossil based) materials that can biodegrade by micro-organisms. For example PBS plastics. This demonstrates that although PBS is not biobased but can biodegrade, it has some association with biology, meaning that it should be included in the definitional model of Biofabricate and thus also in the definition of a 'biomaterial'. But if we have a look at their definitional model, it encapsulates only biobased materials, which PBS is clearly not.

Although, including the 'use' and 'end-of-life' is not sufficient to provide a clearer conception of the term 'biomaterial', since this is commonly used synonymous with 'biobased material'. In addition, it is recognized that the term 'biomaterial' is today associated with the negative connotation of green washing purposes. Meaning applying the term to purposefully make untrue claims about the positive impact a material or product has on the environment. These insights were unexpected and suggest to move beyond the mere materiality of engaging with the term 'biomaterial'. Hence it is advised and argued to be cautious with the term 'biomaterial' and be more specific when talking about materials which are either biobased, biodegradable or both. Furthermore, provided the four identified themes, it makes more sense to elaborate about a process than a materiality. In this respect the term 'biodesign' is introduced in this study. As a result, this study agrees to the definition of 'biodesign' by Biofabricate, reading "A term used to indicate design 'of', 'with' and 'for' biology" [37].

This definition confirms this study's four identified themes as follows.

- Design 'of' biology matches with the material 'origin' and 'production'.
- Design 'with' biology matches with the material 'production' and 'use'.
- Design 'for' biology matches with the material 'use' and 'end-of-life'.

In this way, biodesign utilizes living materials, like plants or cultured tissues, to achieve the goal of organic design, where objects can grow and nature is allowed to take over a portion of the designer's control. In biodesign, nature is often seen as the most powerful, ingenious and scalable technology there is. It has known more iterations to perfect itself, than we can imagine. It recognizes that every aspect inherent to nature lives in a system of its own [47].

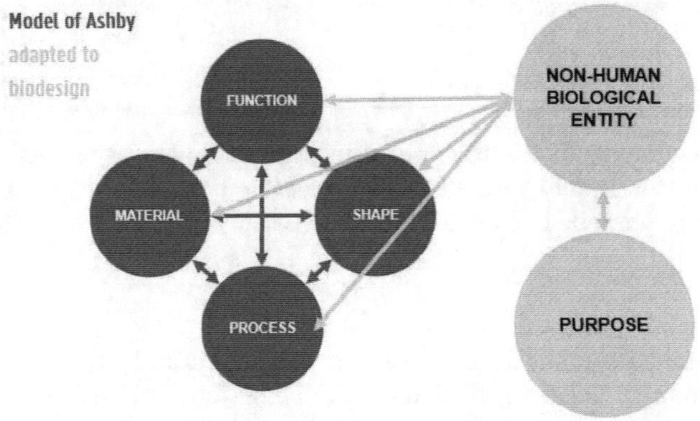

Figure 3. Model of Ashby [48] adapted to the field of biodesign.

As designers we are familiar with the model of Ashby, which depicts the interactions between function, material, process and shape in material selection [48]. However, this study proposes that biodesign goes beyond Ashby's model, as the materials being used are living entities that have an agency and purpose of their own, leading to broader implications (Figure 3). The idea of full control, speed, and repeatability must be revised in biodesign. The central aspect of biodesign is the interplay between human and biological entities, with the latter having some control in the design process. This leads to unpredictability and imperfection, key features of nature that biodesign aims to replicate through collaboration between designers and biologists. This implies that biodesigners are in fact in the last evolutionary role as material 'collaborators'. Collaboration and facilitation between the field of design and the life-sciences, whilst simultaneously collaboration between mankind and nature.

Ultimately, as previously mentioned in the methodology section, a new model to biodesign is suggested that supports categorizing and allocating projects (figure 4) through four lenses; 'origin', 'production', 'use' and 'end-of-life'. Each of these lenses on its own consist of multiple subdomains as categorization criteria (i.e. biobased, biodegradable, biofabrication, bioart, bioreceptivity, biophilia, growing design, and others). It is important to note that the four lenses of biodesign are not restricted to a fixed number of subdomains. Moreover does each subdomain have a definition of its own. This model becomes particularly interesting when trying to assign case-studies to the subdomains of the model. This will often result in having to assign one case-study to multiple biodesign lenses and subdomains. In this way, this model supports in looking at biodesign projects from four different perspectives, but also reflect on each of these lenses how projects impact or disrupt their contextual bio-ecosystems in which they behave. As was stated in the introduction, this contextual reflection is a necessity in order to avoid destructing the bio-ecosystems' core underlying self-sustaining structures.

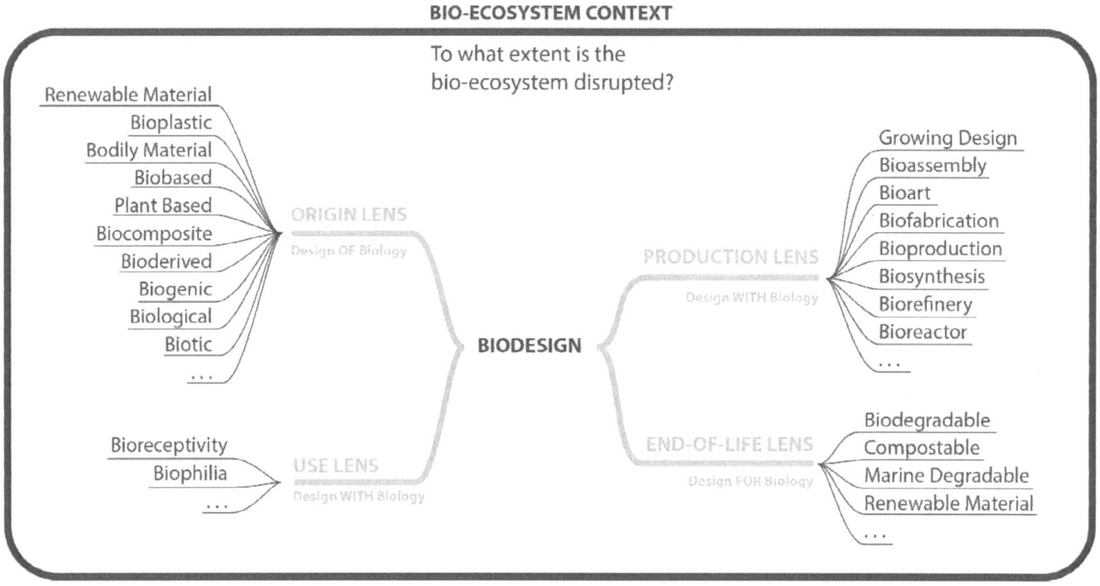

Figure 4. New suggested Biodesign Model comprised of four lenses; origin, production, use and end-of-life. This model supports in looking from different perspectives (lenses) to biodesign projects, including the subdomains it contains (i.e. biobased, biodegradable, bioassembly, and many more open for expansion).

3.2. *Stage 2 – How are 'biomaterials' used today and in future realities*

As has been put forward in the methodology section, is stage two an ongoing research of extensive expert interviews and qualitative analysis to highlight the gaps that currently restricts biodesign from being implemented more in professional design practices. This study would like to stress the importance of performing these dialogues with global experts who are proficient in the field of biodesign. The knowledge, tools, and mindset changes brought about by these individuals' insights, design processes, and experience can hold significant value in spreading awareness about biodesign practices. Therefore, this still ongoing study is a natural consequence of the findings from stage one and previous peer research [26], [34].

While in previous (peer) studies the 'materiality' often dominated the (non-)academic discourse, it is now recognized that actually the 'process' was always the crucial aspect that was at the centre of all discussions. With this rationale, this study hopes to open up the debate on: (1) being more specific in labelling materials (exemplified by ceasing to apply the term 'biomaterial') and (2) talking more about the 'design process' associated with these biological material streams. All of this, originating from this study and the knowledge which is being gained from experts in the field of biodesign.

4. Conclusions

This study set out to explore the intersection of design and biology in today's industry, starting from the ideology behind the continuous evolution of the role of the designer from a materials perspective. The study itself is subdivided into two stages of which the first built upon peer research in this field (such as the suggested pathway to design 'of', 'with' and 'for' biology in an educational context or the implementation of a bio-makerspace in an educational design workspace) by questioning the terminology of and related with the commonly applied term 'biomaterial'. This first stage conducted a broad search over a diverse set of data types with the purpose of documenting bioterms and their definitions. Thereafter the application of a qualitative methodology allowed for comparing and exploring common themes over all the definitions of the bioterms and more specifically for the term 'biomaterial'. The results from stage one suggest to open up the debate about labelling materials, beginning with ceasing the use of the word 'biomaterial' in future studies and thus being more specific about its real properties (i.e. biobased, or biodegradable, or biomanufactured, etc.). In such, this study contributes to enhancing the general understanding of and communication around the biological material streams. A second conclusion that was drawn from stage one, is the recognition of the fact that we are more talking about the 'design process' associated with these biological material streams rather than the pure 'materiality', exemplified by the four identified lenses ('origin', 'production', 'use' and 'end-of-life') in a newly suggested biodesign model. This design process ultimately aligned with the definition of 'biodesign' defined by Biofabricate. Biodesign means to design 'of', 'with' and 'for' biology, aiming at reestablishing a synergetic relationship between mankind and nature. These findings called for stage two, which is still an ongoing research, that aims at uncovering the motives and design processes underlying the biodesign practices of global experts in this field. All of this, with the objective to disclose the gaps which currently restricts biodesign from being implemented in the industry on a professional basis. This ongoing research is performed through conducting extensive expert interviews, a subsequent qualitative analysis and the development of a scheme of concepts that tries to depict the reality of biodesign processes, narrated by experts in the field. The authors of this study plan to disseminate the findings of this ongoing study in a future academic publication.

References

[1] B. Herbert, "Climate Change 2007: The Physical Science Basis," 2007.

[2] G. H. Brundtland, "Our common future - Call for action," 1987. doi: 10.1017/s0376892900016805.

[3] B. Mang, P; Reed, "Update Regenerative Development and Design," in *Encyclopaedia Sustainability Science & Technology*, New York: Springer, 2017.

[4] J. Mensah, "Sustainable development: Meaning, history, principles, pillars, and implications for human action: Literature review," *Cogent Soc. Sci.*, vol. 5, no. 1, Jan. 2019, doi: 10.1080/23311886.2019.1653531.

[5] H. Strömberg, A. Selvefors, and S. Renström, "Mapping out the design opportunities: pathways of sustainable behaviour," *Int. J. Sustain. Eng.*, vol. 8, no. 3, pp. 163–172, May 2015, doi: 10.1080/19397038.2014.1001469.

[6] W. Young, K. Hwang, S. McDonald, and C. J. Oates, "Sustainable consumption: green consumer behaviour when purchasing products," *Sustain. Dev.*, vol. 18, no. 1, p. n/a-n/a, 2009, doi: 10.1002/sd.394.

[7] H. G. Nelson and E. Stolterman, *The design way: Intentional change in an unpredictable world*, Second Edi. Cambridge: MIT Press, 2014.

[8] F. Ceschin and I. Gaziulusoy, "Evolution of design for sustainability: From product design to design for system innovations and transitions," *Des. Stud.*, vol. 47, pp. 118–163, Nov. 2016, doi: 10.1016/j.destud.2016.09.002.

[9] J. Elkington, "Towards the Sustainable Corporation: Win-Win-Win Business Strategies for Sustainable Development," *Calif. Manage. Rev.*, vol. 36, no. 2, pp. 90–100, Jan. 1994, doi: 10.2307/41165746.

[10] B. Giddings, B. Hopwood, and G. O'Brien, "Environment, economy and society: fitting them together into sustainable development," *Sustain. Dev.*, vol. 10, no. 4, pp. 187–196, Nov. 2002, doi: 10.1002/sd.199.

[11] United Nations, "Sustainable Development Goals," 2015. https://sdgs.un.org/goals (accessed Jun. 07, 2023).

[12] F. Ceschin, "How the design of socio-technical experiments can enable radical changes for sustainability," *Int. J. Des.*, vol. 8, no. 3, pp. 1–21, 2014.

[13] M. Saviano, S. Barile, F. Farioli, and F. Orecchini, "Strengthening the science–policy–industry interface for progressing toward sustainability: a systems thinking view," *Sustain. Sci.*, vol. 14, no. 6, pp. 1549–1564, Nov. 2019, doi: 10.1007/s11625-019-00668-x.

[14] N. Cross, "Designerly Ways of Knowing: Design Discipline Versus Design Science," *Des. Issues*, vol. 17, no. 3, pp. 49–55, Jul. 2001, doi: 10.1162/074793601750357196.

[15] H. Rittel, "The state of the art in design methods," *Des. Res. Methods*, vol. 7, no. 2, pp. 143–147, 1973.

[16] P. Vandenbroeck, "Working with wicked problems," *King Baudouin Foundation*. Brussels, pp. 1–50, 2012.

[17] V. Papanek and J. Hennessey, *Nomadic Furniture: How to Build and Where to Buy Lightweight Furniture That Folds, Collapses, Stacks, Knocks-Down, Inflates or Can be Thrown Away and Re-Cycled*. New York, New York, USA: Pantheon Books, 1973.

[18] D. A. Schön, *The reflective practitioner: How professionals think in action*. Basic books, 1984.

[19] European Commission, "Sustainable Product Policy." https://joint-research-centre.ec.europa.eu/scientific-activities-z/sustainable-product-policy_en (accessed Jun. 07, 2023).

[20] A. Waytz and L. Young, "Aversion to playing God and moral condemnation of technology and science," *Philos. Trans. R. Soc. B Biol. Sci.*, vol. 374, no. 1771, 2019, doi: 10.1098/rstb.2018.0041.

[21] J. K. H. Chan, "Design ethics: Reflecting on the ethical dimensions of technology, sustainability, and responsibility in the Anthropocene," *Des. Stud.*, vol. 54, pp. 184–200, Jan. 2018, doi: 10.1016/j.destud.2017.09.005.

[22] L. G. Morris, "The future of biodesign? It's all about collaboration," *Frameweb*, 2020. https://www.frameweb.com/article/sustainability/the-future-of-biodesign-its-all-about-collaboration#:~:text=Biodesign offers incredible means of,way we produce and consume (accessed Feb. 13, 2023).

[23] W. Myers, "The hybrid frontier," in *Biodesign: Nature, Science, Creativity*, Thames & Hudson Ltd, 2018, pp. 8–9.

[24] M. Yadav and M. Agarwal, "Biobased building materials for sustainable future: An overview," *Mater. Today Proc.*, vol. 43, pp. 2895–2902, 2021, doi: 10.1016/j.matpr.2021.01.165.

[25] P. Antonelli, "Vital Design," in *Biodesign: Nature, Science, Creativity*, Thames & Hudson Ltd, 2018, pp. 6–7.

[26] B. Vuylsteke, F. Ostuzzi, J. Detand, and J. Saldien, "Pathway for Designing with New DIY, Circular and Biobased Materials: Insights from Three Case Studies," in *Proceedings of the International Conference on Sustainable Design and Manufacturing*, 2021, pp. 45–54. doi: 10.1007/978-981-16-6128-0_5.

[27] U. Antony and J. Edward, "Flourishing Enterprise Innovation – Tools for the Strongly Sustainable Revolution – Financially Rewarding, Socially Beneficial, Environmentally Regenerative." http://www.flourishingbusiness.org/ (accessed Jun. 19, 2023).

[28] W. Braun, "The system archetypes." p. 27, 2002. [Online]. Available: http://www.albany.edu/faculty/gpr/PAD724/724WebArticles/sys_archetypes.pdf

[29] J. Detand, "Material Oriented Design." 2005.
[30] E. Karana, *Meanings of Materials*. Delft, 2009. [Online]. Available: https://repository.tudelft.nl/islandora/object/uuid:092da92d-437c-47b7-a2f1-b49c93cf2b1e?collection=research
[31] D. L. Kauffman Jr., "Systems One: an Introduction to Systems Thinking," *The Innovative Learning Series*. p. 41, 1980.
[32] S. Parisi, V. Rognoli, and M. Sonneveld, "Material Tinkering. An inspirational approach for experiential learning and envisioning in product design education," *Des. J.*, vol. 20, no. sup1, pp. S1167–S1184, Jul. 2017, doi: 10.1080/14606925.2017.1353059.
[33] Technical Committee : ISO/TC 207/SC 5 Life cycle assessment, "ISO 14040:2006 - Environmental management -- Life cycle assessment -- Principles and framework," 2006. https://www.iso.org/standard/37456.html (accessed Dec. 08, 2018).
[34] B. Vuylsteke, L. Dumon, J. Detand, and F. Ostuzzi, "Creating a Circular Design Workspace: Lessons Learned from Setting up a 'Bio-Makerspace,'" *Sustainability*, vol. 14, no. 4, p. 2229, Feb. 2022, doi: 10.3390/su14042229.
[35] E. D'Itria and C. Colombi, "Biobased Innovation as a Fashion and Textile Design Must: A European Perspective," *Sustainability*, vol. 14, no. 1, p. 570, Jan. 2022, doi: 10.3390/su14010570.
[36] F. Ostuzzi, "Open-Ended Design. Explorative Studies on How to Intentionally Support Change by Designing with Imperfection," 2017.
[37] S. Lee, A. Congdon, G. Parker, and C. Borst, "Understanding 'Bio' Material Innovations: A Primer for the Fashion Industry," 2020. [Online]. Available: https://reports.fashionforgood.com/wp-content/uploads/2020/12/Understanding-Bio-Material-Innovations-Report.pdf%0Ahttps://app.box.com/s/amjq9anszv8hvwdexoxg6wubes4aaxqa
[38] J. (DeZeen) Hahn, "Biomaterials review 2021," 2021. https://www.dezeen.com/2021/12/28/biomaterials-review-2021/ (accessed Feb. 13, 2023).
[39] J. (DeZeen) Parkes, "The Dezeen guide to bio-based materials in architecture, design and interiors," 2021. https://www.dezeen.com/2021/12/09/dezeen-guide-biomaterials-architecture-design-interiors/ (accessed Feb. 13, 2023).
[40] Futurematerialsbank, "Lexicon - Future Materials Bank." https://www.futurematerialsbank.com/lexicon/ (accessed Feb. 13, 2023).
[41] Cambridge Dictionary, "Cambridge Dictionary," 2023. https://dictionary.cambridge.org/dictionary/english/
[42] A. D. Ginsberg and N. Chieza, "Editorial: Other Biological Futures," *J. Des. Sci.*, Sep. 2018, doi: 10.21428/566868b5.
[43] European Environment Agency, "The circular economy and the bioeconomy: partners in sustainability. European Protection Agency Report 8," 2018.
[44] B. Pollini, "Healing materialities," 2022. https://healing-materialities.design/home/ (accessed Feb. 13, 2023).
[45] "Oxford Languages." https://languages.oup.com/ (accessed Feb. 13, 2023).
[46] Wikipedia stub, "Biomaterial." https://en.wikipedia.org/wiki/Bio-based_material (accessed Feb. 13, 2023).
[47] W. Myers, *Biodesign: Nature, Science, Creativity*. Thames & Hudson Ltd, 2018.
[48] F. A. Michael, "Materials selection in mechanical design," *Le J. Phys. IV*, vol. 3, no. C7, 1993.

Research and design for improved packaging of food during daily use outside refrigerated environment

Natalie Waelkens[1], Ben Robaeyst[2], Bastiaan Baccarne[1] & Jelle Saldien[1]

[1] Department Product Development, Faculty of Design Sciences, University of Antwerp, Paardenmarkt 94, 2000 Antwerp

[2] Ghent University, Department of communication sciences, Technicum, T1, Sint-Pietersnieuwstraat 41, 9000 Gent, Belgium

Corresponding author: natalie.waelkens@ugent.be

Abstract. Food waste has a huge impact on economic losses and climate. One of the leading problems is due to lack of effective refrigeration. Transitioning to alternative cooling technologies, such as passive cooling, can offer a potentially more sustainable solution. Therefore, the research investigates how food can be kept cold without relying on a refrigerator or electrical components utilizing the combination of insulation materials and cooling fluid. Using design thinking as a methodology, first of all a user survey was conducted during the study to understand user behavior regarding the storage of chilled food and to define the target audience. Subsequently, an exploratory study was performed to determine the optimal placement of cooling elements to a cooling box for extended cooling. Further, material research was conducted for the inner and outer layer of the cooling container, considering temperature resistance and insulation effects. In parallel with this, an examination was carried out to identify a cooling fluid with efficient cooling capacity and low environmental impact. After implementing this method, it was discovered that the most efficient approach involved placing an ice cooling element at the bottom and around the side walls of a cooling container. Further, the material PLA was demonstrated as suitable to use for cold environments. In consideration of previous research and the application of cooling media in commercial settings, salt solutions were capable of storing food below 4.4 °C. Consequently, the decision was made to proceed with a 20% NaCl solution mixed with water. Ultimately, a final cooling element was designed that fits into the standard IKEA +365 containers measuring 15 x 15 x 7 cm. The cooling capacity of this product was tested and maintained a temperature below 4.4°C for an average of 1.17 hours. This study contributes to existing literature by providing more insights in how passive refrigeration can be aligned with the needs of end-users in order to design both practical and user-fit solutions. However, further research is required regarding the practical implications; to improve insulation, address condensation issues and evaluate compatibility with dishwashers.

Keywords. *Smart packaging, mobile system, user research, food waste, cooling liquid*

1. Introduction

Globally, ⅓ of the edible food produced annually get wasted before it reaches the consumer. This represents about 1.32 tons of food and has economic, ecological, and social consequences. Poor refrigeration is one of the key causes of this issue, leading to food waste throughout the food supply chain. Further, climate change also has an effect on the safety and quality of the food in the cold chain and increases the demand for cooling systems and energy consumption [1-5]. Consequently, temperature abuse occurs in the cold chain, such as storing food at temperatures higher than the optimal storage temperature of 4.4°C, resulting in spoilage, changes in texture, odor, and foodborne diseases [1]. With the world's population expected to surpass 9 billion people by 2050, it is crucial to find innovative solutions for food storage, sustainable cold chains, and changing consumption patterns to prevent resource depletion, reduce carbon footprint, and use resources more efficiently. Proper temperature control throughout the cold chain is essential to ensure food safety and quality [5], [6]. Passive cooling can be proposed as a viable alternative to refrigeration systems due to its low cost, environmentally friendly nature, and energy efficiency. It involves phase-changing materials that can store and release thermal energy without relying on external energy sources [7], [8], [9]. This paper aims to explore new (smart) materials and develop innovative applications for better food packaging in non-refrigerated settings. Using design thinking as a problem-solving methodology, the aim is to tackle the complex challenge of food waste and answer the question of how intelligence can be integrated into food packaging. The end goal is to create a sustainable product that not only offers efficient cooling but contributes to the area of sustainable food storage and consumption practices. With this research, the aim is to gain more scientific knowledge about passive refrigeration and the way food preservation is managed in a rapidly changing world.

2. Methodology

Design thinking is a powerful methodology that plays an important role in the design process. It can be divided into five parts [10]. The first phase is the empathy stage, where one seeks to understand the needs and perspective of the user. Next, defining where the problem is clearly defined. Further, in the ideate phase, variations of solutions are generated. In the fourth stage prototype, a rough representation of the solution is created. In the test phase, the prototype will be evaluated and refined based on user feedback [11].

2.1 Empathize and define

The objective is to visualize the parties involved in the product by creating a stakeholder map. Furthermore, the aim is to identify the target group more precisely through a survey to gain insights into the behavior and habits of individuals who transport chilled food during their daily commute. A more elaborate view on the selected target groups can be found in the result section. In addition, in-depth interviews were conducted with the target group to find a connection between their behavior, routine, and the obstacles they encounter while commuting. By combining the survey data and findings from the interviews, it is possible to construct a representation of how consumers handle food in realistic scenarios, the types of food they bring, and personas can be developed [12]. Following this, a competitor analysis was conducted on ten different cooling boxes, evaluating them based on various criteria such as weight, volume, grip capacity, etc. Spider diagrams were utilized to visualize the performance of these cooling boxes and identify potential opportunities for designing improved alternatives. Furthermore, in order to gain a thorough understanding of user behavior and their effort to maintain the temperature low of their food, journey maps were generated [13]. These maps provide insights into the various interaction's users have with their food throughout the day. Subsequently, a function tree is formulated, highlighting the essential functions that must be incorporated in the design of a cooling box. To ensure a comprehensive exploration of design possibilities, a morphological map is then developed. This map offers a diverse array of combinations, facilitating the generation of a wide range of innovative concepts.

2.2 Ideate

In the second part, various concepts were generated and developed based on the insights from the previous phase. To select the most user-friendly concepts, a day-in-life story was created from the perspective of the personas. Each concept was presented in detail to a user to understand how it would function in a real-life scenario. Additionally, users were asked additional questions using the think aloud protocol to capture important remarks and feedback from users. To make the collected data more tangible and assist in decision-making, a trade-off matrix was created. Four users were asked (N=4) to assign weights to nine design criteria. By analyzing this matrix, the best concept for further development can be determined. Furthermore, feedback was sought from stakeholders such as IKEA and Euro point to gather input and make decisions regarding the design direction.

2.3 Prototype

Based on the results obtained so far, a test was conducted to determine the cooling capacity of a standard square IKEA +365 container with dimensions of 15x15x7 cm. During this test, 400 ml of water is added to the container, which is then placed in a refrigerator overnight to simulate daily conditions. Four different test setups were performed to determine the optimal placement of the cooling element: 1. Cooling element inside the container at the bottom, 2. Cooling element outside the container at the bottom, 3. Cooling element at the bottom of the lid inside the container, and 4. Cooling element upright in the middle of the container. Each of these tests were conducted with a cooling element stored in both the refrigerator and freezer. In total, nine tests were conducted, including one test without a cooling element as a reference. Using a DS18b20 temperature sensor and an Arduino Uno, temperature was measured and monitored in real-time for four hours [14]. The goal of this test is to investigate the optimal placement of the cooling element in a lunchbox [15]. Subsequently, based on the results of the cooling experiments, new concepts were formulated, and a trade-off matrix was generated with the input of four users. Then, a material investigation was conducted to determine the suitable material for the inner and outer layers of the cooling element [16]. The most environmentally friendly refrigerant that can provide sufficient cooling power is also examined. To determine the materials for the inner and outer layers, Edupack is used to compare properties such as thermal conductivity and minimum user temperature. Additionally, infill percentages for 3D printing were examined to add additional insulating properties to the material by creating an insulating air layer [17]. Furthermore, a decision is made regarding the refrigerant to be used in the prototype. By researching commercially used refrigerants, the refrigerant with good cooling performance, suitability for temperatures below 4.4 °C, and environmental impact is determined [18] . Factors such as the chemical and physical properties of the solution, including its impact on corrosion, were taken into consideration. Calculations were performed to determine the required salt concentration for creating a phase change material (PCM) solution, as described in equation (1).

$$\Delta Tf = Kf * n == Kf * \frac{m}{M}$$

Equation 1: Freezing Point Depression

In this equation, ΔT_f represents the freezing point depression, K_f is the molar freezing point depression constant, n is the number of moles, m is the mass, and M is the molar mass. For water, K_f is equal to 1.86 °C.kg/mol, which means that 1 mole freezes at -1.86 °C instead of 0.00 °C [19]. Lastly, the determination of the thickness of the cooling element is briefly discussed by considering the heat transfer between the cooling element, the environment, and the food. Using thermodynamic formulas, the total heat transfer of the cooling element is calculated as described in equations (2), (3), and (4).

$$Qtotal = Qwalls + top + Qcooling\ liquid + Qbotton$$

Equation 2: Total heat transfer of the lunch container

$$Q = \frac{(T2 - T1)}{((RA + RB + RC + RD))}$$

Equation 3: Heat transfer by conduction

$$R = \frac{d}{k.A}$$

Equation 4: Thermal resistance of a material

Here, Q represents the heat transfer through the four materials in W, R_A to R_D represent the thermal resistance of each material in K/W, T_1 represents the temperature inside the lunchbox (4.4 °C), T_2 represents the ambient temperature (21°C), k represents the thermal conductivity of a material in W/m.K and A represents the surface area through which heat flows in m^2. To complete these calculations, the thermal conductivity of the chosen cooling fluid also needs to be determined. This value is obtained from a graph based on research conducted by Thermtest Instrument™ and averages at 0.55915 [20]. Finally, using Siemens NX software, a prototype is developed, which is then 3D printed and filled with the appropriate cooling material.

2.4 Test

Once the prototype is designed, a general evaluation is made, providing an overview of the final product appearance and its practical use in a real-life environment. To further assess its performance, a scenario will be simulated using a storyboard, allowing for a recreated experience with the product. To survey users' perceptions of the prototype, written feedback will be collected, and noteworthy comments carefully documented. This was performed with seven participants (N=7). Subsequently, the prototype will undergo an analysis by repeating the previously conducted cooling test. This evaluation aims to identify any differences with previous setups and measure the effectiveness of the prototype in maintaining the desired temperatures of 4.4 °C. Furthermore, the use of a thermal camera will allow visualizations to be made, enabling the observation of temperature variations in the final product for four hours. Moreover, a study will be conducted on the compatibility of the prototype using it in a dishwasher, examining whether the prototype can withstand cleaning in a dishwasher environment. Finally, a separate section will address the sustainability aspect of the product, examining its environmental impact and sustainability characteristics.

3. Results & discussion

3.1 Survey

The response to the survey included 43 participants from Belgium with a mix of age groups, the majority of whom were between 21 and 25 years old. Specifically, 22% of the respondents were students, while the majority (73%) were working. In terms of socio-economic settings, among the working individuals, a great amount of people works in sectors such as finance, healthcare and IT. Among the students, the majority of the students were pursuing academic studies, while others were enrolled in professional programs. Regarding transportation, 62.5% of the students primarily used bicycles, while 92.6% of the working individuals used cars for their commute. The study examines the types of food the participants carried with them. The majority (78%) indicated that they carried perishable foods that required refrigeration. The main reason for bringing food was to consume it during lunchtime. In terms of food packaging, 34 participants used lunchboxes, while 10 participants used plastic or aluminum containers. Furthermore, it was found that 17.6% of the participants used ice packs to keep their food fresh when refrigeration facilities were not available. On the other hand, 29.4% kept their food at room temperature in their bags throughout the day.

As a result, 37% of the participants discarded their food upon returning home, either because it had not been kept refrigerated or because it no longer seemed appetizing. To understand the relationship between modes of transportation and food waste, the percentage of participants who disposed of their food was analyzed. Out of the 16 participants who disposed of their food, because it had not been refrigerated for an extended period, one-third were students who primarily traveled by bike, while two-thirds were working individuals who primarily used cars. For this group, it is challenging to keep their lunch cool throughout the day as they often lack access to a refrigerator or other cooling options while on the go. Based on these findings, the target audience is divided into two categories: students who travel by bike and employees who travel by car.

3.2 In-depth interviews

To gain a better understanding of user habits, six focus interviews were conducted with respondents to establish a connection between their behavior, routine, and the barriers they face in storing chilled food. The research revealed that consumers do not pay particular attention to storing chilled food they transport from home to work or school. Several users kept perishable foods at room temperature for the entire morning due to the unavailability of a refrigerator. Moreover, most lunches were not transported in an insulated cooler box or with cooling packaging.

3.3 Competitor analysis

Subsequently, a benchmark study was conducted, which was visualized using spider diagrams. All spider diagrams were overlaid to identify areas for potential improvement. The results are presented in Figure (1).

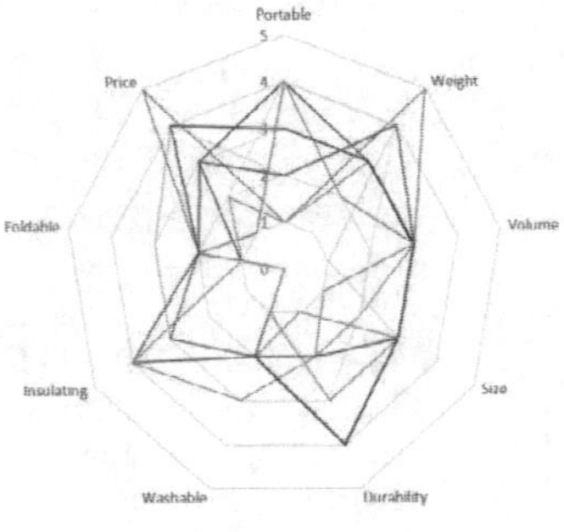

Figure 1: Spider diagram of all benchmark products

The conclusion was that the main features that could be adjusted were the volume and size of the lunchbox. Many of the existing products were found to be excessively large in size. Additionally, the washability of the lunchbox was an important point. The compared products were difficult to clean or couldn't be put in the washing machine. Therefore, in the design process, it will be important to minimize the weight and overall dimensions of the lunchbox. Furthermore, the development of a box or ice pack that can be placed in the dishwasher would be a valuable improvement.

3.4 Journey maps

After analyzing the results of the benchmark study, the next step is to investigate how users handle their food on a daily basis. For this purpose, a schedule was created to depict the daily routine of both students and working individuals, including their commuting times and when they typically consume their meals. An assumed schedule is created, along with estimates of how long food is exposed to ambient temperature. This amounted to at least 5.25-6 hours from 7:00 a.m. to 1:00 p.m. And if they don't consume their meal in the afternoon, this can extend to 10 to 13 hours. It is important to note that this is just an estimate based on travel times and does not account for unexpected delays or other factors that could prolong the time without refrigeration. More user tests need to be conducted to obtain more accurate results. In addition to the limitations, the schedule provides a general idea of how long food remains unrefrigerated in a backpack. By gaining insight into the issues users face on a daily basis, designs can be created that better meet consumer needs, thereby improving the user experience and the quality of the food. Incorporating these findings into the design requirements ensures that the refrigerated packaging effectively addresses the challenges users face and delivers optimal refrigeration performance for food preservation. The design requirements include aspects such as weight, volume, size, durability, washability, insulation, foldability, price, reusability, and cooling efficiency.

3.5 Ideate

Based on the function tree and morphological map, 8 concepts were generated, out of these, 3 concepts were excluded due to the potential integration into the other five concepts. Proceeding with the selected 5 concepts, a "day in the life" scenario is written, explaining to users how the product would work in practice. It was found that some people have no issue leaving their lunch unrefrigerated in their bag throughout the day, while others were aware of potential changes in taste and concerns about food safety. It is proposed to work on an adaptable design that can accommodate different shapes and sizes, utilizing standard containers that users already have at home. The ice packs can be adjusted according to the required cooling capacity. An overview of the total weight provided by 4 users for the 5 different concepts is shown in Figure 2.

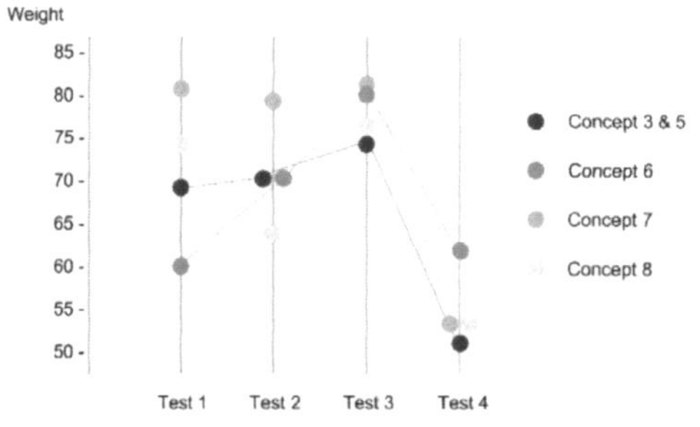

Figure 2: Overview total weight given by four users to five different concepts

Based on these results, it was decided to proceed with concept 7. This decision was based on the findings from the user tests and the information obtained from IKEA and Euro point. It was decided to create an ice pack for an existing lunchbox, such as the standard containers from IKEA. This company already has a modular system where each lid fits different boxes of the same size. It is interesting to develop an ice pack for this purpose, creating a modular cooling system. This approach offers more flexibility so that consumers can continue using their existing containers and only need to add an ice pack. This also reduces waste since there is no need to produce new lunchboxes.

3.6 Prototype

To determine the position at which an ice pack can keep the food cool for the longest duration, temperature measurements were conducted at various positions of the ice packs. The results of these measurements can be found in Table 1.

Table 1: Overview results temperature cooling capacity test

	Min. T(°C)	T after 2 hours (°C)	T after 4 hours (°C)
Position C1	2.50	10.50	14.63
Position C2	5.75	12.50	15.63
Position C3	5.13	12.06	16.44
Position C4	4.88	15.63	19.94
Position F1	1.37	3.44	5.81
Position F2	5.50	9.13	10.94
Position F3	3.25	7.44	9.25
Position F4	2.13	6.13	10.69
Position R	6.45	13.75	18.44

After two or four hours, all positions exceeded the recommended temperature of 4.4 degrees Celsius, except for test F1 where the ice was placed at the bottom of the lunchbox. However, the minimum temperature of the water in test F1 was 1.37 degrees Celsius, which should also be considered in terms of potential effects on taste change. A similar study that has been conducted by Du et al. in 2020, examining the placement of phase-changing panels in a cooler box to generate the longest cooling time. It confirmed that placing the cooling around the box and one at the bottom resulted in the longest cooling time of 9.50 hours [9]. Therefore, the decision was made to proceed with frozen food coolers, as they have greater cooling capacity and longer durability compared to those from the refrigerator. Additionally, placing an ice pack inside a lunchbox with pre-existing food is not user-friendly, which led to the decision to pursue an ice pack on the outside of the lunchbox. Based on Du et al.'s research, it was also decided to include additional ice packs on the sides for extra cooling. However, it is important to note that the available data is incomplete and insufficient for drawing accurate conclusions. Various parameters, such as time, initial temperature, water volume, ambient temperature, and seasonal variations, need to be taken into consideration, and more experiments should be conducted to obtain more reliable results.

Based on this exploratory study, new concepts were generated, and feedback was collected using a trade-off matrix to determine which concept aligns best with the ultimate goal. Concept 7 emerged as the most favorable, and based on that, a 3D model was created in Siemens NX, consisting of three main components. The first part encloses the IKEA +365 box, the second part is placed at a distance of 1 cm to create a cavity between the two parts, and the third component is used to seal the top of the cavity and contain the coolant inside.

Subsequently, a material investigation was carried out. Firstly, the material for the inner and outer layers of the ice pack was determined. Using Edupack, a graph was created considering thermal conductivity and the minimum temperature at which it can operate [21]. This allowed the selection of four polymers: Polyethylene (PE), Polylactide (PLA), Polyvinyl chloride (PVC), and Polypropylene (PP). PVC can withstand temperatures up to -18.2 °C, which falls just outside the range for a freezer. Moreover, PE and PP were found to be less recyclable compared to PLA. Therefore, the decision was made to proceed with the biodegradable polyester PLA, as it is more environmentally friendly and suitable for 3D printing. This material will be used for both the inner and outer layers of the ice pack. To maximize the insulating properties of the outer layer, it was also decided to increase the thickness of the outer wall compared to the inner wall by using a 10% infill percentage during 3D printing. This adds an additional layer of air, which has an insulating effect and slows down heat transfer.

Furthermore, it is important to make a good choice for the coolant of the lunch containers. Several articles have discussed water mixtures with inorganic salts as efficient phase-change materials. These materials are inexpensive and readily available. Eutectic salt-water solutions have a high energy storage density and are within the melting range suitable for cooling food below 4.4 °C. The thermal properties of salt solutions, such as thermal stability, were examined. The article by Yang et al. revealed that solutions of NH4Cl and NaCl remain stable for 150 minutes. As the salt concentration increases, the phase transition period will be longer, which is favorable for food cooling [22]. In terms of physical properties, it is also necessary to consider volume changes during the freezing of the coolant. When water freezes, it expands. By adding salts to the water, the hydrogen-oxygen bonds are disrupted, resulting in less expansion of water during freezing. In terms of kinetic properties, the phenomenon of undercooling occurs when the liquid struggles to freeze as the temperature drops below the melting point. Therefore, it is necessary to significantly lower the temperature below this point. The study by Yilmaz et al. focused on NaCl and KCl, indicating that a NaCl salt solution causes less undercooling than a KCl solution. Particularly, concentrations of 20% NaCl showed the least undercooling [23, 24]. Furthermore, consideration was given to possible chemical reactions between the outer and inner layers of the ice pack, such as corrosion. The effect of corrosion on commonly used materials has already been tested in various articles. Based on the study by Oro et al., it was concluded that polymers such as polypropylene (PP), polyethylene terephthalate (PET), high-density polyethylene (HDPE), and polystyrene (PS) did not suffer damage when exposed to different salt solutions [25]. Subsequently, the freezing point was calculated for the 20% NaCl-water solution using the formula for freezing point depression (1). Adding salt lowers the freezing point. The freezing point depression for the 20% NaCl solution amounts to 12.73 °C. Ultimately, for the design of the end product, this cooling liquid of 20% NaCl solution was used. Additionally, the thickness of the ice pack was calculated based on heat conduction formulas. The results obtained from filling in formulas (2), (3), and (4) are presented in Table 2, along with the thickness of the material.

Table 2: Overview results heat transfer

	Q (W)	Material thickness (mm)
Four sides of the lunchbox in PP	31.27	3
Outer layer in PLA	16.01	4
	8.00	8
Inner layer in PLA	12.74	2
+ PP layer		3
Top lid in PP	22.74	3
Outer layer PLA	93.22 – 151.92	4 – 8
+ cooling media		6 – 2
+ inner layer PLA		2 - 2
+ PP layer inside		3 - 3

From this, it can be inferred that a greater thickness of the outer layer results in a larger insulating effect because there will be less heat transfer. Additionally, a thick layer of coolant reduces heat transfer. It is chosen to proceed with a thickness of 4 mm for the outer layer and a thickness of 4 mm for the coolant. It is important to note that these estimates are incomplete and do not account for numerous other factors that can affect the performance of the ice pack. For instance, different types of food have varying thermal conductivity, which can impact the rate at which they cool. Furthermore, temperature fluctuations due to seasonality, temperature variations within the refrigerator, and temperature transfer from the lid to the food can further complicate the calculations. To obtain more accurate estimates, it is necessary to utilize specialized software that takes all these parameters into account to calculate the thickness of the ice pack. This is beyond the scope of the research but may be interesting to consider for future calculations.

3.7 Test

The final design was tested with the target audience to gather feedback on its usability. The participants were asked to provide written opinions based on a storyboard depicting the user context. The feedback revealed that the product was perceived as useful because it can keep food cool, is compact and portable, but is only compatible with IKEA +365 containers. One of the respondents that commutes every day to his job observation was:

"It is possible to use this item for different containers. For example, you still need to wash your cooling element from Monday and another cooling element can be placed in the freezer. Then on Tuesday you can use the cooling element with another container."

Further there were concerns about potential long-term changes in shape that could make the product fit less well with an IKEA container. Furthermore, condensation occurs during food cooling, necessitating further improvements and tests regarding condensation, wear and tear, and wash ability. Subsequently, the same cooling capacity test was performed twice on the final concept to assess its cooling performance. The same protocol as the previous temperature tests was followed, and the results are presented in Figure 3.

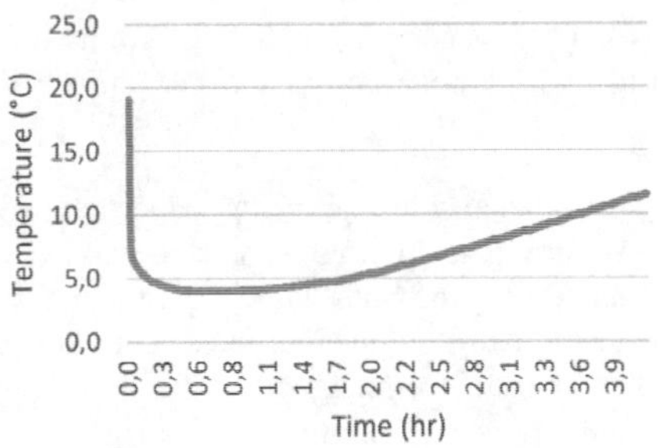

Figure 3: Result final product temperature in function of time, position E1

In Table 3, the values of this test were compared with those of test F3, where an ice pack was placed outside a lunchbox. When using a PCM as a cooling agent, the temperature remains stable below 4.4 °C for an average of 1.17 hours before gradually increasing. In test F3, the temperature rises much faster once the minimum temperature is reached. Over a 4-hour period, test F3 manages to stay below 10 degrees, while the PCM exceeds this threshold. Graphically, F3 exhibits a concave shape, in contrast to the convex shape of the PCM graph.

Table 3: Overview results cooling capacity test final product

	Time x < 4.4 °C	Time x < 7 °C	Time x < 10 °c	After 4 hours
Position E1	1.44	2.62	3.65	11.06
Position E2	0.89	2.03	3.21	11.31
Position F3	0.51	1.74	-	9.25

Based on thermal camera images, a visual representation is shown of the cooling element to assess how the cold spreads across the cooling element over a period of 4 hours. It is evident that the coldest part is located on the inside of the cooling element and becomes warmer towards the outside. After four hours, the temperature has risen from -12.1 to 17.0 degrees. Additionally, the occurrence of condensation remains a challenge when using the product as it can moisten the inside of backpacks. Further research needs to be focused on improving insulation. Furthermore, the possibility of using the product in a dishwasher was considered. Material research indicates that PLA has limitations in terms of its maximum temperature range. A dishwasher operates between 48 and 72°C, while PLA has a maximum operating value of only 54.9°C. Thus, PP and PE are considered more suitable materials. Additionally, further research must be conducted on the cooling performance of the cooling medium after undergoing different temperature cycles, which may vary in duration. Ultimately, a test was conducted in a dishwasher where a tear in the inner PLA layer caused leakage, leading to salt crystallization. Further research should be carried out to improve these areas.

4. Conclusion and discussion

Keeping food cool for a minimum of 4 hours is a challenging issue. The research involved four phases to develop a definitive cooling product through user research. The target audience consists of cycling students and commuting employees, which is consistent with the majority of the survey participants, who mainly travel by bicycle or car. Through routine and behavioral research, including surveys and interviews, it was determined that this target group often does not pay attention to properly cooling perishable food. By conducting cooling capacity tests on food containers with different placements of the cooling element, it was determined that using a freezing element at the bottom and sides of a cooler box is the most suitable alternative cooling method. Through material research, a product made of PLA was developed, employing various thicknesses to optimize insulation. An environmentally friendly cooling medium based on a 20% NaCl solution was chosen as an alternative to more traditional and toxic cooling agents. The final product was validated through the same cooling capacity test, which showed that working with the PCM used in the product results in a more stable and slower temperature rise below 7°C compared to using a more traditional ice pack. This design not only provides insulation for maintaining food freshness but also contributes to reducing food waste. By keeping the temperature stable and under a certain point of 4.4 degrees, the product helps to extend the shelf life of perishable items, reducing the likelihood of food spoilage. However, there are still some side effects that require further investigation, such as condensation, improved insulation capabilities, and possibilities for dishwasher resistance. By addressing these concerns and continually striving for improvement, the product has the potential to make a positive impact on reducing food waste and promoting sustainable consumption practices.

When comparing this research to other technologies, this paper contributes to the exploration of sustainable solutions to food waste. by examining technologies such as passive cooling that can be environmentally friendly alternatives to traditional cooling methods while meeting end-user needs. By analyzing user behavior, solutions are sought that are adapted to the specific selected target groups. In addition, material research ensures the selection of suitable materials for refrigeration containers, with a focus on temperature resistance and insulation. The efficiency and environmental impact of refrigerants are taken into account. Looking at the implementation of intelligent technologies, there is ongoing research on IoT devices and mobile apps that help reduce food waste. By incorporating intelligence into these solutions, potential environmental benefits can be achieved. Passive cooling is one of them and represents a promising area of research with considerable potential.Consequently, with this paper we aim to provide research that can be situated on the nexus of sustainable food waste solutions, user-fit products and intelligent technologies.

5. References

[1] 2013 Food and Agriculture Organization of the United Nations. and Food Wastage Footprint (Project) *Food wastage footprint : impacts on natural resources : summary report* FAO

[2] B. Lipinski 2013 *By the Numbers: Reducing Food Loss and Waste*

[3] 2023 World Wildlife Fund. WWF & 1986 Panda Symbol, "Food Waste Overview." Accessed: Mar. 15, 2023. [Online]. Available: worldwildlife.org

[4] L. Xu, D. W. Sun, Y. Tian, L. Sun, T. Fan, and Z. Zhu Oct 2022 ACS Appl Mater Interfaces *Combined Effects of Radiative and Evaporative Cooling on Fruit Preservation under Solar Radiation: Sunburn Resistance and Temperature Stabilization (*vol 14 no 40) doi: 10.1021/acsami.2c11349 pp 45788–45799

[5] 2022 *Sustainable food cold chains: Opportunities, challenges and the way forward* UNEP FAO doi: 10.4060/cc0923en

[6] I. - International Energy Agency, "The Future of Cooling Opportunities for energy-efficient air conditioning Together Secure Sustainable." [Online]. Available: www.iea.org/t&c/

[7] L. Otter Products, "PASSIVE VERSUS ACTIVE COOLING," 2021. Accessed: May 20, 2023. [Online]. Available: www.drawdown.org/solutions/refrigerant-management

[8] I. Oropeza-Perez and P. A. Østergaard 2018 Renewable and Sustainable Energy Reviews *Active and passive cooling methods for dwellings: A review* (Elsevier Ltd vol 82) doi: 10.1016/j.rser.2017.09.059 pp 531–544

[9] J. Du, B. Nie, Y. Zhang, Z. Du, li Wang, and Y. Ding J Apr 2020 Energy Storage *Cooling performance of a thermal energy storage-based portable box for cold chain applications (*vol 28) doi: 10.1016/j.est.2020.101238.

[10] Philippe Vandenbroeck, Working With Wicked Problems. Brussels: The King Baudouin Foundation, 2012. [Online]. Available: www.kbs-frb.be

[11] R. Wolniak May 2017 *The Design thinking method and its stages*

[12] S. Balzan, L. Fasolato, B. Cardazzo, G. Berti, and E. Novelli Dec. 2014 *Cold chain and consumers' practices: Exploratory results of focus group interviews* (Ital J Food Saf vol 3) doi: 10.4081/ijfs.2014.4516 no 4 pp 241– 244

[13] S. Mercier, S. Villeneuve, M. Mondor, and I. Uysal July 2017 *Time–Temperature Management Along the Food Cold Chain: A Review of Recent Developments* (Compr Rev Food Sci Food Saf vol 16 no 4) doi: 10.1111/1541-4337.12269 pp 647–667

[14] Safe Food Australia, "2-hour/ 4-hour rule," InfoBite.

[15] J. Du, B. Nie, Y. Zhang, Z. Du, li Wang, and Y. Ding Apr 2020 *Cooling performance of a thermal energy storage-based portable box for cold chain applications* (J Energy Storage vol 28) doi: 10.1016/j.est.2020.101238

[16] 2017 Cambridge: Granta, "CES EduPack."

[17] Y. Y. Aw, C. K. Yeoh, M. A. Idris, P. L. Teh, K. A. Hamzah, and S. A. Sazali Mar 2018 *Effect of printing parameters on tensile, dynamic mechanical, and thermoelectric properties of FDM 3D printed CABS/ZnO composites* (Materials vol 11 no 4) doi: 10.3390/ma11040466.

[18] S. Yilmaz, F. Sheth Karathia, I. Martorell, H. O. Paksoy, and L. F. Cabeza *Salt-water solutions as PCM for cooling applications*
[19] J. W. Hedenquist and R. W. Henley 1985 Economic Geology *The importance of CO2 on freezing point measurements of fluid: evidence from active geothermal systems and implications for epithermal ore deposition* (vol 80 no 5) doi: 10.2113/gsecongeo.80.5.1379 pp 1379–1406
[20] Thermtest Instrument TM 2023 Thermtest Instruments TM *Thermal Conductivity of Salty Water Mixtures (Water + NaCl)* Thermtest.com
[21] Cambridge: Granta, "CES EduPack." 2017.
[22] E. Oró, L. Miró, C. Barreneche, I. Martorell, M. M. Farid, and L. F. Cabeza 2013 *Corrosion of metal and polymer containers for use in PCM cold storage* (Appl Energy vol 109) pp 449–453 doi: 10.1016/j.apenergy.2012.10.049
[23] Huseyin Ozbek and Sidney L. Phillips *Thermal conductivity of aqueous NaCl solutions*
[24] S. Yilmaz, F. Sheth Karathia, I. Martorell, H. O. Paksoy, and L. F. Cabeza *SALT-WATER SOLUTIONS AS PCM FOR COOLING APPLICATIONS*
[25] E. Oró, L. Miró, C. Barreneche, I. Martorell, M. M. Farid, and L. F. Cabeza, "*Corrosion of metal and polymer containers for use in PCM cold storage* (Appl Energy vol 109) pp 449–453, 2013, doi: 10.1016/j.apenergy.2012.10.049.

Educational Board Game Design Based on Local Wisdom for Youth to Support Social Sustainability Learning

Melita Mulyani and Johanna Renny Octavia

Department of Industrial Engineering, Faculty of Industrial Technology, Parahyangan Catholic University, Ciumbuleuit 94, Bandung 40141, West Java, Indonesia

melitamulyani0605@gmail.com and johanna@unpar.ac.id

Abstract. Social sustainability is the process of creating sustainable, successful places to offer prosperity by understanding what people need from where they live and work. In the 17 goals of the SDGs, social sustainability is reflected in the goals of the pillars of social development which include the goals of (1) no poverty, (2) no hunger, (3) healthy and prosperous life, (4) quality education, (5) gender equality. Adolescents who are the next generation must also begin to know the principles related to social sustainability. Besides, it is proven that people who play board games can absorb and memorize material more effectively than conventional learning techniques. This research resulted an educational board game design that collaborates on local wisdom performances from West Java as an adaptation of the Pagelaran Yogyakarta board game to generate revenue for solve social problems. The results obtained from the testing were an increase in the post-test score to 83.33 from the initial pre-test value of 52.38 and the average final result of the questionnaire on sustainability awareness was 3.19 from the initial average of 2.56. This indicates an increase in sustainability awareness, especially regarding social sustainability for adolescents from moderate to high levels.

1. Introduction

Today, the world is moving rapidly towards change in many things like digitization, globalization, and the degradation of the natural world. With such a big change, the issue of sustainability is something that is being pressed socially, economically, and environmentally[19]. According to Fischhoff, Agarwal, and Gilvesy (2021) in their article entitled What is Social Sustainability, sustainability is generally defined as something that is done to meet the needs of the present without sacrificing the ability of future generations to meet future needs. In 2015, the United Nations set 17 sustainable development goals, also known as the Sustainable Development Goals, with 169 targets aimed at measuring the social, economic, and environmental dimensions of sustainability adopted by 193 countries with a mission to them by 2030[22].

Social sustainability is one of the aspects of sustainability both in the concept of the triple bottom line and in the goals and objectives of the Sustainable Development Goals. Social sustainability is the process of creating a sustainable place, a place that succeeds in offering well-being by understanding what people need from where they live and work[20]. In the 17 goals and 169 SDGs, social sustainability is described in the social development pillar goals that include (1) no poverty, (2) no hunger, (3) healthy and prosperous living, (4) quality education, (5) gender equality[22]. Social sustainability occurs when processes, systems, structures, and formal and informal relationships actively support the capacity of future generations to create healthy and habitable communities. A sustainable social community that is just, diverse, connected, and democratic and provides a good quality of life[1].

The concept of social sustainability is often associated with urban life, because as time goes by, there are many problems within urban areas, especially its inhabitants. Designer of the concept of social sustainability design, revealed that now and in the future, housing is becoming crucial given that people are in great need of it, especially those in urban areas given the large number of migrations to urban areas[21]. If this context is linked to social sustainability, then social sustainability is the link between physical world design and social world design which means that although housing is important, aspects such as social relationships, social networks, and cultural development must also be taken into account. Social sustainability must also ensure that physical infrastructure and social life run side by side where in such a unity there is room for society to develop, form groups and young people who are the seed of future generations can grow well. So from that, it takes the role of the government to understand how important it is that society not only needs building physically, but also knows how to serve and encourage people to improve their quality of life in society.

Not only the government, the youth who are the seed of the next generation must also begin to learn about two principles related to social sustainability, namely social justice and sustainability of society[6]. Social justice refers to the definition that every individual has equal opportunities to engage in economic, social, and political activities. Moreover, these young people are expected to begin to understand that each individual also has an equal opportunity to access public service, education or training. When linked to the goals of the social development pillar in the SDGs, youth will begin to be introduced to various issues of social development that need to be addressed such as no poverty, no hunger, health and well-being, gender equality, and education. Meanwhile, the concept of community sustainability is a principle that emphasizes urban communities to strengthen their communities both now and in the future[6]. From there, the young as a successor generation must gradually begin to become aware of the concept of social sustainability and also begin to understand the principles related to social justice and the sustainability of today's societies so that in the future, they can still maintain these principles so that social sustainability can be created.

The preliminary interview study was conducted with 6 respondents with the criteria of 15-19 years of age. Based on the interview, it was identified that 5 out of 6 adolescents who became respondents did not know about the concept of sustainability, especially the concepts of social sustainability. Jean Piaget, in his theory entitled Piaget's Stages of Cognitive Development, identifies the four stages of cognitive development of children, one of them is about the formal operation for adolescents aged 11 and over with the development of the main characteristics of hypothesis, abstract, deductive and inductive, as well as logical and probability[14].

Hansmann (2010) in his writing on Sustainability Learning, argues that affective and cognitive learning outcomes serve as input variables for the next sustainability learning process. Maslow (1954) understood motivation as the process by which individuals respond to a need by doing something to fulfill it. Maslow assumed that humans are constantly motivated by a need, if a certain need is satisfied, individuals become motivated to satisfy other needs. This is demonstrated in the theory of needs, the Maslow Triangle, which consists of a hierarchical need model that distinguishes between physiological needs, needs for security, social needs, self-esteem needs, and needs for self-actualization. Maslow's need classification has been widely recognized and has proven useful in developing indicators of social well-being that cover important aspects of social sustainability[4].

The conceptual distinction between intrinsic and extrinsic motivation is essential to understanding the learning process. Intrinsic motivation can come from curiosity and interest in the learning activity itself, such as the excitement felt when analyzing, asking, exercising, and solving problems[11].

In addition to the problem of the concept of social sustainability that is not much known to teenagers, on the other hand, it turns out there is also the issue of local wisdom that is beginning to be marginalized. Local wisdom loaded with messages, values, cultures, and character passed down from generation to generation has been packed in various forms such as regional songs, regional dances, regional arts, local food, even to local games. As the times evolve, there will be a tendency of local wisdom that is loaded with the various cultural values that are beginning to be marginalized. So one way to keep this local wisdom alive and there is to socialize it to the younger generation so that this local Wisdom does not disappear as time goes by. Dwijendra (2020) in his research proves that the presence of local wisdom has a significant factor in shaping urban patterns and structures. The results of Sukawati research (2017) also show that the values of local wisdom of the community always accompanied the development of a region from the beginning of civilization to become an international tourist area.

A preliminary interview study was conducted with the same six respondents in the previous interview with the criteria of 15-19 year-olds who aimed to know the current state of their knowledge about the local wisdom of West Java. Based on the interview, it was identified that 6 of the 6 respondents said that during this time they knew about the kinds of local Wisdom West Java only through cultural arts lessons on the school bench but limited only to some kind of Local Wisdom such as Angklung and some local songs such as Manuk Dadali and Cingcangkeling. Preliminary studies of literary searches were also carried out and it was found that the awareness of Sundays to consciously preserve their culture was very low. This is proved by about 500 kinds of Sunda's art that are almost extinct because there is no regeneration of its players. The true attitude of the good people has been very difficult to find in the daily attitudes of the Sunda society of the present day[8].

One method that can be used for social sustainability learning and also local wisdom is through board games. Board games are a space for learning mathematics and a learning space that enables learning a variety of content. Board games enable a range of interactions that involve players in computational thinking, teamwork, and creativity[2]. The relationship between board games and learning has been proven across disciplines and states. Board games simplify complex problems and systems, which makes it suitable to further explore learning and concepts such as motivation and computational thinking in formal and informal settings[2].

Noda, Shirotsuki, & Nakao (2019) in their research showed that board games and programs that use board games have a positive effect on a variety of outcomes, including educational knowledge and cognitive function. Board games are an effective tool to encourage active learning and knowledge retention and can help increase student motivation to learn and even lead to positive behavioral change[13]. Sucianto (2019), based on a summary of 83 different studies from different countries on board games that have been studied over the past seven years, states that people who play board games or obtain gamified learning materials have been shown to absorb and memorize material more effectively than conventional learning techniques.

Based on the all explanations before, the aim of this study is to design a local intelligence-based educational board game for adolescents as a social sustainability learning advocate that will address the problem of under-knowledge of social sustainable concepts especially by adolescents and as one of the ways for teenagers to continue to know and learn local intelligences. In addition, it will also evaluate the design of board games based on field testing results and find out the comparison of sustainability awareness teenagers before and after playing the designed board games.

2. Research methods

The research was carried out in several stages as can be seen in Figure 1.

Figure 1. Research methods.

2.1. *Identification of needs*

The first step is to identify the needs of the board game design. The identification of the needs is done to find the requirements of the type of board game, the local wisdom, and the concept of social sustainability. Interviews and observations are the methods used to identify those needs. The first interview was done to find the need for the kind of board game. The interview was conducted with six teenagers, a project-based learning educational practitioner and an educational game creator, and four master games. Based on the results of the interview in the form of a transcript of each of the respondents, the process of identification of customer statements that indicates the needs of the consumer is carried out. In the process of interpreting consumer statements, there are some guidelines to follow that need statements are statements "what" and not "how" the product works, writing statements must always be positive rather than negative, should not use "should" and "better", the needs presented are attributes of the product not from the user side, and must be specific. Examples of interpretation of a need statement of a consumer statement can be seen in Table 1.

Table 1. Example interpretation of the need for board game type based on the statement of adolescent respondents 1.

Customer statement	Need statement
"It's confusing when you're in a game, but you usually don't know how to play it. But if you like it, you'll always like it. It's always a game and it's not gonna make a lot of boredom."	A board game that's easy to understand.

The interview process was discontinued when no new requirements were found from the last three sources. This can happen because both latent and hidden needs have been identified so that if further interviews are conducted on a new source, the needs expressed will be the same or similar to the needs previously obtained. Then from that, on the production of cumulative needs data, the cumulative of the identified needs statements are sought. The amount of needs in the next consumption sequence that is calculated is just different needs and not identified in the previous consumer. This phase is repeated until no new requirements are obtained on at least the last 3 sources.

Observation methods are also used to identify the needs of board games. Observations are carried out at one of the cafe board games in Bandung which aims to observe how interactions are created between players during playing the board games they play, what kind of board game they like, as well as what board games that they don't like. The observations are done to 7 groups of teenagers, consisting of a group of teens who are playing a board game and also the group of adolescents who are looking for a game to play.

Based on the interpretation of the needs of all respondents interviewed as well as the results of observations, there are some kinds of needs that have something in common. From that, a grouping of consumer needs into several groups of needs is done by looking at the commonalities of different needs using affinity diagrams. The data used is the need statement data obtained at the data collection stage using both interview methods and observation methods. The obtained needs are then separated and placed in groups already defined based on the similarity of the need statement.

Based on the grouping of needs that have been done, then 9 groups of needs are produced. Re-interviews were conducted to the same respondents to give an assessment of the existing needs group with the following range of assessment, unwanted (undesirable) with 1 point, not important with 2 points, would be nice to have, but isn't necessary with 3 points, highly desirable, but I would consider a product without it with a value of 4 points, and critical feature with a value of 5 points. This is the result of a list of needs sorted by the most important levels: board games that interesting, board games that easy-to-learn and understand, boards games with interesting type of interaction, board games that easily carried everywhere with simple and durable components, board games with family type game, board game for 4-8 players, board games for 2-10 players, board games with abstract strategy type games, and board games with short duration of play. Based on the assessment of the board game needs groups, the needs about board games for 2-10 players were eliminated because the needs board games for 4-8 players had a higher average value. It's the same thing with the needs about board game with abstract strategy game type which were also eliminated as board games with the family game type had higher average values. A group of board game needs with a short duration of play was also eliminated as the majority of respondents did not worry too much about the duration.

In the same way but with local wisdom's respondents, here's a list of needs sorted by the most important levels: board game that can introduce West Javanese art performances, board game that can introduce traditional weapons from West Java, board game that can introduce traditional games from

West Java, board game that can introduce traditional ceremonies from West Java, board game that can introduce traditional songs from West Java, board game that can introduce traditional dances from West Java, board game that can introduce traditional dress from West Java, board game that can introduce traditional musical instrument from West Java, board game that can introduce traditional foods from West Java.

Last, still with the same way but with social sustainability's respondents, here's a list of needs sorted by the most important levels: a board game that contains the concept of sustainability which can be learned through small habits that can train responsibility in daily activity; a board game that can teach how to end hunger, achieve food security and good nutrition, and promote sustainable agriculture; a board game that can teach how to create inclusive and equitable quality education and increase lifelong learning opportunities for all; a board game that contains the concept of sustainability which can be learned through a lifestyle; a board game that contains the concept of sustainability which can be learned through a variety of sustainability lifestyles, a board game that can teach how to end poverty in all its forms everywhere; a board game that can teach how to achieve gender equality and women empowerment; a board game that contains the concept of sustainability which can be understood by youth about concept of social sustainability; a board game that can teach how to create a healthy life and improve the welfare of all ages.

2.2. Concept development

The development of alternative board game concepts begins with the development of implementation plans for each list of existing needs. For example, plans for the implementation of local wisdom content needs are structured in the form of local types of wisdom according to identified needs. Meanwhile, plans for implementing social sustainability learning needs are constructed in the shape of implementation plans through game mechanisms. The next step is to perform benchmarking through some surveys as well as searching for previously existing board games whether they are board games on local intelligence or board games about learning social sustainability. The benchmark results and things adapted for board game design can be seen in Appendix A.

Based on the results of the drafting of the implementation plan and the benchmarking results obtained, the main sketch of the educational board game is formed based on local wisdom to support learning of social sustainability. The main sketch basically has two main concepts in it, namely the concept of local Wisdom that is widely adapted from the Pagelaran Yogyakarta board game and the social sustainability concept. Using the SCAMPER method and from the benchmarking results, the main sketch has been developed into four alternative sketches. Overall, concepts 1 and 2 have similarities with concepts 3 and 4 in the form of the game board layout. However, for game mechanisms, concept 1 has a similarity with concept 3 and concept 2 has a common nature with concept 4.

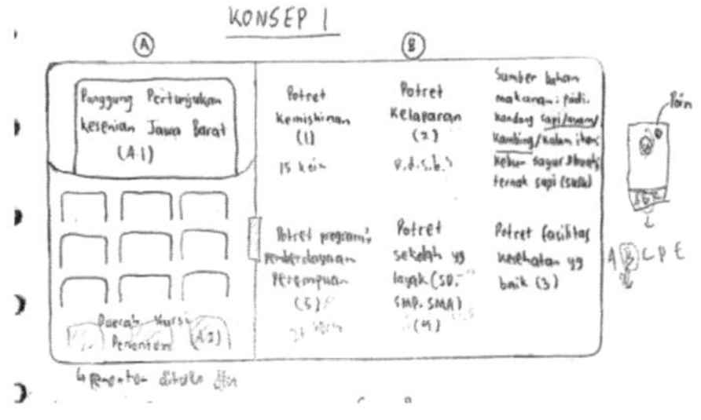

Figure 2. Concept 1.

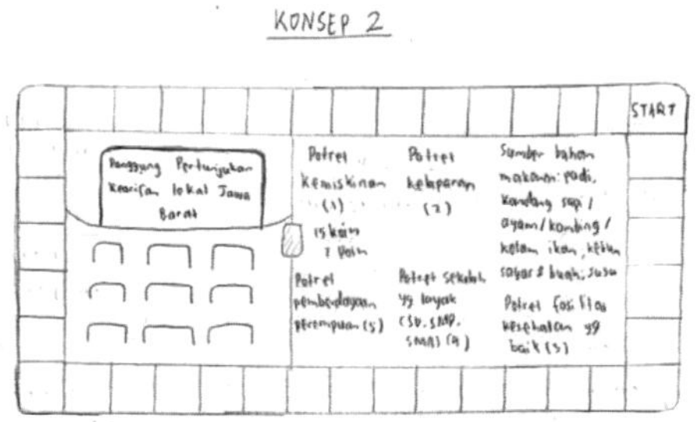

Figure 3. Concept 2.

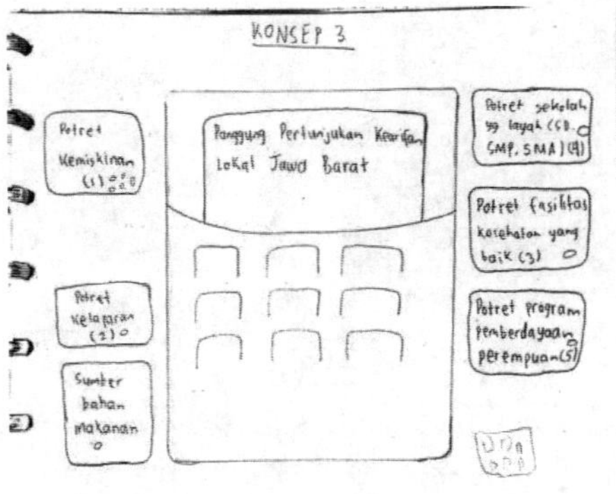

Figure 4. Concept 3.

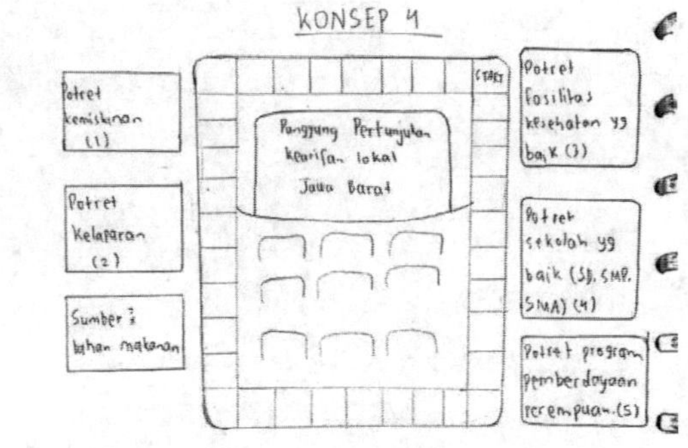

Figure 5. Concept 4.

2.3. Concept selection

Concept selection is done by external decision method which means that the four existing alternative concepts are determined by an external entity which in this case is a teenager aged 15-19. The concept selection process was carried out by interviewing six teenagers who also became the source of the interview at the identification stage. Four out of six adolescents interviewed chose alternative concept 1 while there were two teenagers choosing alternative concept 4. So, the concept 1 was chosen to be developed at the prototype stage.

2.4. Prototyping

All data that has been generated from the previous phase is used in the prototyping phase. In this phase, paper prototypes and interactive prototype are used. Paper prototypes or alpha prototypes are formed with the aim of ensuring that the mechanisms of the board game are running properly and all the content that is to be delivered through this board game is delivered. From the test results using the paper prototype, there are some improvements implemented, one of which is about pricing and the impact that occurs when a social problem has been successfully solved. This can be seen in Table 2.

Table 2. Costs and impacts for every social problem.

Social problem	Cost	Benefit
Poverty	24 coins	Discount 3 coins for the cost of the other problems
Hunger	20 coins	+3 spectator capacities inside the show building
Healthy and prosperous life	16 coins	+1 coins/spectator
Quality of education	12 coins	+ 1 card slot in hand
Gender equality	8 coins	+1 coin/spectator, when there is an actress on the card that being played

The next step is to create an interactive prototype. The main board can be seen in Figure 6, while the other components can be seen in Figure 7 and Figure 8.

IOP Conf. Series: Earth and Environmental Science 1256 (2023) 012006 doi:10.1088/1755-1315/1256/1/012006

Figure 6. Main board game.

Figure 7. Local wisdom card (1/60).

Figure 8. Spectator cards, player's pawn, coins, and first player token.

2.5. Testing

The board game test was conducted in two sessions, the first being conducted on two teenagers aged 15-19 and the second on four 15-19-year-olds. They are the same respondents as the respondents involved from the identification stage of the need. The game board testing process is performed by playing the game from start to finish. The documentation of the testing process can be seen in Figure 9 and Figure 10. After the test phase is completed, the stage continues with the board game evaluation.

Figure 9. First session of board game testing.

Figure 10. Second session of board game testing.

2.6. Evaluation

Board game evaluation is carried out using three methods, there are post-play interviews, pre-test and post-test, consisting of 14 multiple-choice questions about social sustainability knowledge that was provided by the board games, and a pre-sustainability awareness-questionnaire and a post-sustainability awareness-questionnaire. Interviews are conducted to find out about users experience when playing the board game.

Board game evaluation is also carried out using pre-test and post-test questions consisting of 14 double-choice questions. Frisbie (2005) stated that questions in the form of double-choice have better reliability and validity compared to the type of true question. Furthermore, the issues used are largely guided by metadata of social indicators on the pillars of social sustainability development under the SDGs, namely Objectives 1 to Objective 5. The creation of a double choice has followed good steps according to Brame (2013).

Referring to the study of Noordin T A and Sulaiman S (2010) there are three categories of sustainability awareness namely sustainability practice awareness, behavioral and attitude awareness, and emotional awareness. The concept of determination of such assessment instruments was adopted from the Saptaji et al. (2020) research on the development of instruments to measure the sustainability awareness of high school students on temperature and caloric materials. The validity and reliability of the assessment instrument has been tested. From the results of the validity test, there are a number of invalid assessment points, so that the valid assessment details are not adopted in the formulation of the questionnaire. The results will be processed using the average data referring to the level of sustained awareness by Nuroso & Rini (2022).

3. Results

Based on the interviews carried out, here are some of the evaluation results as well as suggestions for improvement given by the respondents: there are still some local wisdom cards with images that do not match the name of the local Wisdom, the price to solve the problem of poverty and hunger is too expensive, first player tokens are not very useful, the number of local Wisdom cards should be multiplied again because when played by 4-5 people will run out of the stock of cards, the material to make the player's character stand should be stronger so that the character does not fall when played. In addition to the evaluation and suggestions for improvement, there are also some new things learned or known by the respondents after playing the board game, one of them said that this board game has the potential to be developed to schools for the lessons of Cultural Arts or can be played when no classes.

Based on the recapitulation of pre-test and post-test results, it can be seen that there is an increase in the value from pre-test to post-tests of each respondent. This can also be seen from the average pre-test value that previously had a score of 52.38 has increased to 83.33 in the after-test value. Meanwhile, sustainability awareness results before playing board games were 2.56 in the medium category, while sustainability awareness outcomes after playing the board game were 3.19 in the high category. This indicates that the board games designed can raise the social sustainability consciousness of teenagers from the medium to the high level.

The evaluation of sustainability awareness through this sustainability awareness questionnaire is only used to measure the personal awareness of each teenager based on what they should do from the questionnaire after playing a board game. However, to continue to maintain, nurture, and develop a consciousness of social sustainability must also be supported by the wide range of knowledge they have acquired about social sustainability. This awareness of social sustainability can also continue to be built through daily behavior so that the practice of values and concepts of social sustainability can continue to flourish and live within each adolescent.

4. Conclusions

The study concluded that local wisdom-based educational board games for adolescents to support social sustainability learning were designed to collaborate on the concept of local wisdoms showing in West Java that was adapted from Yogyakarta's Pagelaran board games to generate income that would be used to solve various existing social problems. There is an advantage that will be gained by each player after solving each social problem to motivate the players in solving the social problem. On the other hand, the proposed improvement of the prototype of the board game based on the testing of prototypes suggests that there is still a need to make some improvements to the prototype that has been made, among other things, the standee material for the player's character has been further strengthened so that the character can stand well and exchanged the impact of the poverty problem when it is successfully resolved with the effect of the health and well-being problems when it has been successfully solved.

For further research, it would be better for the values of each type of local wisdom to be more detail so that they are more relevant to the topic of social sustainability that is being taught. Hopefully, the values present in the local wisdom can be implemented and live from generation to generation in realizing social sustainability in the surrounding environment.

Appendix A

Table A1. Benchmarking.

Name	Objective	Components	Pictures	Things adapted
The Festivals	Promote tourism potential in Indonesia.	Game board, player marker, block, each player's cover box	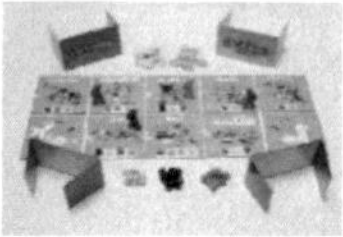	Marker's player
Stone Age	Leading the ancient human tribe so that it can survive like fit plants, produce new generations, etc.	Game boards, dice, player markers, goal tiles, etc.	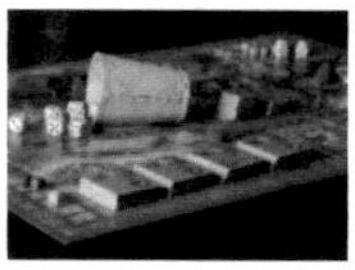	Social problem content
Pagelaran Yogyakarta	Saving the stage from bankruptcy by collecting 50 ratings in five months so the stage doesn't close.	Game board, cards, coin tokens, player markers		Mechanisms and games related to local wisdom
Masospoli	Problem Based Learning as an alternative medium in sociology learning.	Game boards, cards, dice		Education concept of board games
Petualanga n Pensi	Implants social behavior according to Pancasila's values in pre-school children.	Game board, touchscreen monitor & panel, android TV box, speaker, two-hole socket, challenge card		Education concept of board games
Keliling Kota	Introducing traffic alarms to 5-6-year-olds.	Game board and cards	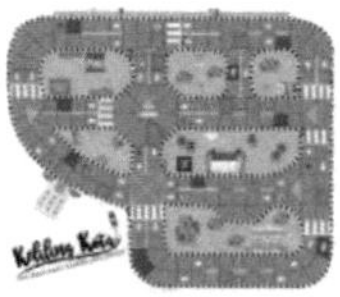	Education concept of board games
Gejala Depresi	It communicates the symptoms of depression to the general public so that it can be more sensitive to depression and helps prevent the possibility of increased depression.	Cards		Education concept of board games

References

[1] Barron L and Gauntlet E 2002 *WACOSS housing and sustainable communities indicators project.* In Sustaining our Communities Int Local Agenda 21 Conf, Adelaide (pp. 3-6)

[2] Bayeck R Y 2020 *Examining Board Gameplay and Learning: A Multidisciplinary Review of Recent Research.* Simulation & Gaming, **51(4)**, 411–431. https://doi.org/10.1177/1046878119901286

[3] Brame C 2013 *Writing good multiple choice test questions* Retrieved [todaysdate] from https://cft.vanderbilt.edu/guides-sub-pages/writing-good-multiple-choice-test-questions/

[4] Clarke M, Islam S M N, and Paech S 2006 *Measuring Australia's well-being using hierarchical needs.* Journal of Socio-economics, **35(6)**, 933–945. https://doi.org/10.1016/j.socec.2005.11.063

[5] Coursera 2023 *What Is Sustainability? (And Why It's Important)* Retrieved from https://coursera.org/share/412ac563d31ef0be90d295eb8239f373

[6] Dempsey N Brown C and Bramley G 2012 *'The key to sustainable urban development in UK cities? The influence of density on social sustainability'*, Progress in Planning, vol. **77**, no. 3, pp. 89-141. https://doi.org/10.1016/j.progress.2012.01.001

[7] Dwijendra N K A 2020 *How the Local Wisdom Influences the Sustainability of Spatial Development in Denpasar, Bali, Indonesia* Bhumi : Jurnal Agraria Dan Pertanahan https://doi.org/10.31292/jb.v5i3.389

[8] Editorial 2013 *Kurang, Kesadaran Masyarakat Sunda Menjaga Budaya Sunda - Universitas Padjadjaran* Retrieved from https://www.unpad.ac.id/2013/01/kurang-kesadaran-masyarakat-sunda-menjaga-budaya-sunda/

[9] Fischhoff M, Agarwal D, and Gilvesy J 2021 *What is Social Sustainability?* Retrieved from https://nbs.net/what-is-social-sustainability/

[10] Frisbie D D 2005 *MULTIPLE CHOICE VERSUS TRUE-FALSE: a COMPARISON OF RELIABILITIES AND CONCURRENT VALIDITIES* Journal of Educational Measurement, **10(4)**, 297–304 https://doi.org/10.1111/j.1745-3984.1973.tb00807.

[11] Hansmann, R 2010 *"Sustainability Learning": An Introduction to the Concept and Its Motivational Aspects* Sustainability, **2(9)**, 2873–2897. https://doi.org/10.3390/su2092873

[12] Maslow A H 1954 *Motivation and Personality*; Harper & Brothers: New York, NY, USA.

[13] Noda S, Shirotsuki K, and Nakao M 2019 *The effectiveness of intervention with board games: A systematic review* Biopsychosocial Medicine, **13**. https://doi.org/10.1186/s13030-019-0164-1

[14] Pakpahan F and Saragih M 2022 *Theory Of Cognitive Development By Jean Piaget*. Journal of Applied Linguistics **2** 55-60. 10.52622/joal.v2i2.79.

[15] Rini N and Nuroso H 2022 *PROFIL SUSTAINABILITY AWARENESS SISWA SMA/SMK PADA MATERI SUHU DAN ENERGI.* Jurnal Sains Dan Pendidikan Fisika, **18(1)**, 68. https://doi.org/10.35580/jspf.v18i1.21535

[16] Saptaji A E, Chandra D T, and Wijaya A 2020 *Pengembangan Instrumen untuk Mengukur Sustainability Awareness Siswa SMA pada Materi Suhu dan Kalor*. Seminar Nasional Fisika, **1(1)**, 11–21. Retrieved from http://proceedings.upi.edu/index.php/sinafi/article/download/1242/1121

[17] Sucianto K, 2019 *4 Alasan Bermain Board Game Baik Buat Kesehatan Mental Kamu* Retrieved from https://www.intothelightid.org/2019/09/26/4-alasan-bermain-board-game-baik-buat-kesehatan-mental-kamu/

[18] Sukawati T 2017 *Establishing Local Wisdom Values to Develop Sustainable Competitiveness Excellence.*

[19] Trisakti, MM CSR 2018 *Kenapa Keberlanjutan Penting untuk Masa Depan?* Retrieved from https://majalahcsr.id/kenapa-keberlanjutan-penting-untuk-masa-depan/

[20] Woodcraft S 2011 *Design for Social Sustainability: A Framework for Creating Thriving Communities*. London:The Young Foundation.

[21] Woodcraft S 2012 *Social Sustainability and New Communities: Moving from Concept to Practice*

in the UK. Procedia - Social and Behavioral Sciences, **68**, 29–42. https://doi.org/10.1016/j.sbspro.2012.12.204

[22] Yulaswati V, Primana J R, Oktorialdi, Wati, D S, Maliki, Moeljono A N S.,.... Buana E C 2020 *Metadata Indikator Tujuan Pembangunan Berkelanjutan (TPB)/Sustainable Development Goals (SDGs) Indonesia Pilar Pembangunan Sosial*. Retrieved from https://sdgs.bappenas.go.id/website/wp-ntent/uploads/2020/10/Metadata-Pilar-Sosial-Edisi-II.pdf

IOP Conf. Series: Earth and Environmental Science 1256 (2023) 012007 doi:10.1088/1755-1315/1256/1/012007

Enhancing West Java's Horticultural Transportation: Design Thinking for Food Loss Reduction

Sandra Helenna Mulyadi, Carles Sitompul, Johanna Renny Octavia

Industrial Engineering Department, Parahyangan Catholic University, Jl. Ciumbuleuit 94 Bandung 40141 Indonesia

Corresponding author: sandramulyadi19@gmail.com

Abstract. The purpose of this paper is to reduce food loss during transportation from farmers to post harvester in West Java, Indonesia. Agriculture is one of the sectors that support the country's economy and horticultural commodities, especially vegetable and fruit products, are very susceptible to both damages caused by pests and post-harvest activities. The design thinking methodology is applied to reduce food loss for banana transportation from farmers to post-harvester, which include the cycle of activities of emphatize – define – ideate – prototype – test. A transportation handling is designed for transporting banana in an attempt to reduce food loss.

1. Introduction

Indonesia is an agricultural nation because the majority of its citizens are employed in the agricultural sector, which supports the country's economy. Indonesia's agro-climatic and agroecosystem conditions are highly conducive to the development of diverse horticultural production due to the year-round availability of sufficient sunlight, the varying elevations of sea levels, and the varying temperatures and humidity. Horticulture is the cultivation of garden plants, including fruits, vegetables, ornamental plants, and medicinal plants. As each region in Indonesia has a different potential for producing vegetables and fruits, it is possible to fulfill the needs of each region by distributing supplies from other regions. Horticultural products, particularly vegetable and fruit products, are extremely susceptible to physiological and mechanical damage from pests and post-harvest activities. The transportation process is crucial to the supply chain because it is the key to profitability and affects the total cost of the supply chain. Shocks in the transportation process can result in losses, a reduction in product quality, which is a form of mechanical damage to horticultural commodities that refers to rot. Generally, frictional damage consists of scrapes, abrasions, and cracks.

Consequently, a decline in the quality of horticultural products can result in food loss and waste. Food waste is a food loss that occurs in the retail and final consumption stages due to the behavior of retailers and consumers, i.e., discarding food. Food loss is a decrease in the mass of edible food in all stages of the supply chain, especially those leading to food that can be consumed by humans, namely losses in the post-harvest and processing stages. Consequently, research is required to reduce food loss for horticultural commodities that are susceptible to diminished product quality as a result of common mechanical damage during transportation.

Consideration must be given to the decline in product quality of horticultural commodities along the supply chain, particularly during the transportation process. This is because these factors can increase the company's total expenses, while reducing revenues. By considering these factors, businesses can make more informed decisions regarding their supply chain. The Indonesian Directorate General of Horticulture cultivates 323 types of commodities, including 60 types of fruits, 80 types of vegetables, 66 types of medicinal plant commodities, and 117 types of floriculture commodities. The production of horticultural products in Indonesia exceeds 10 thousand tons, and cabbage is one of the vegetables with the highest harvest quantity, while bananas are one of the fruits with the highest harvest quantity [1]. Initially, observations and interviews were conducted at the *Caringin* and *Gede Bage* Markets, the two largest traditional markets in Bandung, the capital of West Java Province, Indonesia. Observations and interviews with market traders revealed that a significant amount of food loss (organic waste) occurs daily. Particularly organic waste resulting from post-harvest processing when horticultural products reach the traditional market.

2. Introduction

Horticultural commodities, particularly vegetable and fruit products, are extremely vulnerable to physiological and mechanical damage from pests and post-harvest activities [2]. Post-harvest handling, such as storage, packaging, and transport, is crucial for maintaining product quality [2]. Figure 1 shows some of the mechanical damages on tomatoes, which includes bruises, scratches, and cracks.

Figure 1. Mechanical Damage on Tomatoes [1]

At the start of the research, observations and interviews took place at three wholesale markets in Bandung: Pasar Andir, Caringin, and Ciroyom, on March 5, 2021. These activities revealed a significant amount of horticultural waste, particularly cabbage vegetables. Observations and interviews with market traders were conducted to collect information. From these interviews with seven traders, it was found that a bag of cabbage vegetables typically weighs 50-60 kg, with about 10 kg of outer leaves discarded due to poor quality. In cases of exceptionally bad quality, up to 10-15 kg could be discarded from one bag. The interviews also uncovered that vegetable waste from the market is usually disposed of at a waste disposal site, known as the final disposal site (TPA). In Pasar Caringin, approximately 15 trucks of organic waste, each carrying 17 tons, were generated daily. This waste can be categorised into two types: unsold and rotten products and products becoming waste before being sold due to prior processes.

A supply chain is an integrated business process made up of all the components directly and indirectly involved in meeting consumer demand [3]. The distribution process is crucial to the supply chain because it is the key to profitability and affects the total cost of the supply chain [3]. Transportation is defined as the movement of goods from one location to another, beginning at the origin of the supply chain and ending with the consumer. Transportation typically comprises between one-third and two-thirds of a company's total logistics expenses [2]. The green supply chain is more profitable than the conventional one [4]. Traditional supply chains aim to reduce costs and increase supply chain efficiency in order to maximize economic returns. Green Supply Chain Management (GSCM) is implemented to improve environmental performance, which is calculated by reducing air emissions, effluent waste (liquid waste from toilets or kitchens), solid waste, and toxic material consumption [5]. When supply chains generate waste, expenses rise [6]. Eliminating waste sources will help reduce costs, both calculated costs, such as the cost of waste disposal, and unaccounted costs, such as the workers' quality of life. Design thinking is a problem-solving strategy utilized to address "wicked problems." A wicked problem is a complicated problem with no obvious solution. Food loss is a difficult problem because it is obscure, persistent, and pervasive [7]. In addition, the issue of food waste, which is food loss, has an effect on resources such as water, energy, land, and labour, as well as a similar effect on the environment as food waste, which produces methane gas that can cause a greenhouse effect. Design thinking can be used to reduce food loss in the supply chain, particularly between farmers and traders of horticultural commodities, in order to address this issue. Consequently, the following is the formulation of the problem in this study: How can design thinking be used to reduce food loss during the transportation of horticultural commodities between farmers and traders?

3. Design Thinking Methodology

In this study, a supply chain model was developed to map the parties in the supply chain. Furthermore, the determination of the limitation of the problem is carried out marked with a dotted line box. The supply chain modelling aims to map the supply chain state of horticultural commodities according to field conditions.

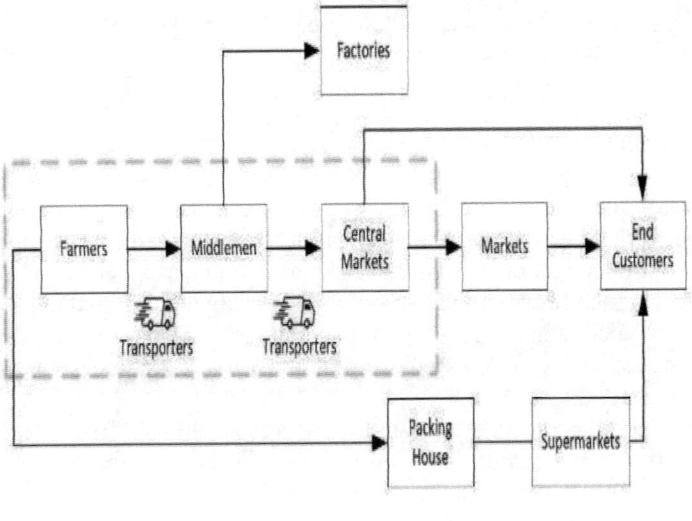

Figure 2. Horticultural Supply Chain Model

This research methodology contains the steps taken in carrying out this research process. Thus, the research methodology is useful to facilitate and make the research process more systematic and structured. In conducting this research, a research methodology is needed. Design thinking methodology consists of several phases, which can be described as follows.

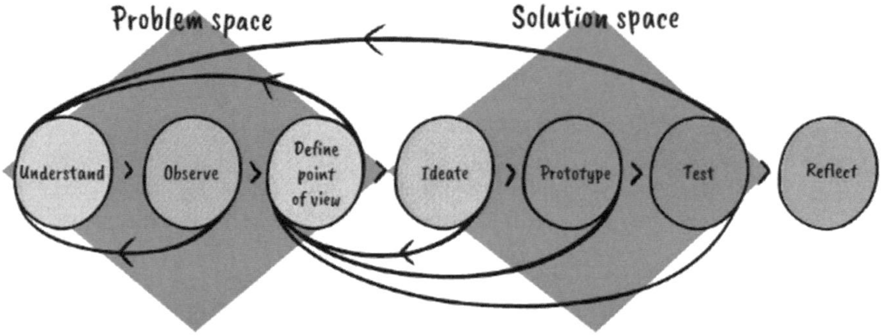

Figure 3. Design Thinking Methodology [8]

Empathy activities are carried out to gain an understanding of the problem to be solved. Then, a problem question formulation is made. This phase aims to **understand** as much as possible about the needs of potential users. Interviews were conducted for the empathy process. To **observe** is to learn as much as possible about users and their needs. Then, recap the results of the interview process with the aim of describing the results of the interview clearly. Next, can determine the point of view. **Define Point of View** is the end of the problem analysis. The problem formulation is carried out as a point of view that will be used later as a starting point for finding solutions to problems. Problem formulation was carried out using the "How might we..." questions. **Solution Ideas** are made to solve existing problems by involving parties related to the research. Generating ideas is done as much as possible. Next, the best idea is selected from the ideas that have been created using the 2x2 Matrix tool. The **prototype** of the selected idea is then created. The low-fidelity prototype was made and subsequently continued with a high-fidelity prototype. Prototyping is a manifestation of the ideas created. Finally, **testing** on the prototype is executed with the possibility to refine the solution of the problem. Figure 3 shows the design thinking methodology.

4. Results and Discussion

The supply chain of horticultural products, particularly cabbage and bananas, begins with farmers who play a role in the planting and harvesting of products. Then, farmers distribute their harvests to intermediaries (to point of collection). From intermediaries, some products are distributed to factories, while others are distributed to (large) market traders. Traders are those who purchase large quantities of goods directly from manufacturers for resale to retailers or industrial firms (big market traders are often known as traders). The modelling of the supply chain aims to map the supply chain status of horticultural products based on field conditions.

The process is then continued with the phases of comprehend and observe. The application of design thinking begins with the understanding phase of empathy. Every participant in the supply chain, from farmers to merchants, engages in a process of empathy to comprehend the problem. Farmers, middlemen, transporters (drivers), and market traders are the existing parties (large). The understanding stage is conducted using semi-structured interviews, specifically exploratory interview tools and empathy interviews. Next, persona profiles for farmers, intermediaries, transporters, and market traders are created. A summary of the results of the empathy stage was then provided.

Table 1. Empathy Result (Cabbage)

Cabbage			
No.	Causes Food Loss	The Party Experiencing the Loss	% from Quantity
1	Bad weather (High rainfall.)	Farmers & Middlemen	± 0% - 30%
2	Cabbage quality is poor due to pests.	Middlemen	± 2% - 5%
3	Decreased quality of cabbage due to friction in post-harvest handling processes.	Farmers dan Middlemen	± 20% - 30%

Table 2. Empathy Result (Banana)

Banana			
No.	Causes Food Loss	The Party Experiencing the Loss	% from Quantity
1	Early harvest.	Farmers and Middlemen	± 2% - 3%
2	Decreased quality of bananas from bunchers / mechanical damage to bananas due to friction in post-harvest handling processes.	Farmers and Middlemen	± 10% - 25%

In addition, the process of determining the point of view is used to formulate a question that will aid the subsequent phase, which is the ideation phase for solutions to existing problems. To determine the perspective, the "How might we..." question is utilized. Two questions are determined based on the formulation of the questions to determine the point of view. How can we prevent cabbage and banana damage from farmers to traders? and how can we prevent banana damage from farmers to traders? After determining the perspective, we proceed to the solution's ideation phase.

As inspiration for solution creation, analogies and benchmarking are used as tools. These tools are used to generate ideas through an exploratory approach to contexts beyond the problem statement. In addition, brainwriting, a tool for solution generation, is employed. Using the brainwriting technique, seven people contributed to the conception of solutions. Then, 28 solutions for the cabbage problem and 24 solutions for the banana problem were generated. We then prioritize the implementation of ideas in the first quadrant, namely those with a high impact and a low level of difficulty, as determined by the mapping results.

Then, two concepts for cabbage were obtained, namely cabbage containers and handling instructions. In addition, two ideas were obtained for bananas, namely banana containers and banana-handling instructions. After selecting ideas from solution ideation results, prototypes are created. For the cabbage and banana container concept, prototyping begins with the creation of a low-fidelity prototype in the form of a simple two-dimensional illustration created with paper, drawing pens, and colour tools. Using

SolidWorks, the researcher then created a high-fidelity prototype. To determine the dimensions of the container, data on the types and dimensions of vehicles commonly used to transport horticultural commodities as well as the dimensions of horticultural commodities were collected. Figure 4 depicts a low-fidelity cabbage container prototype.

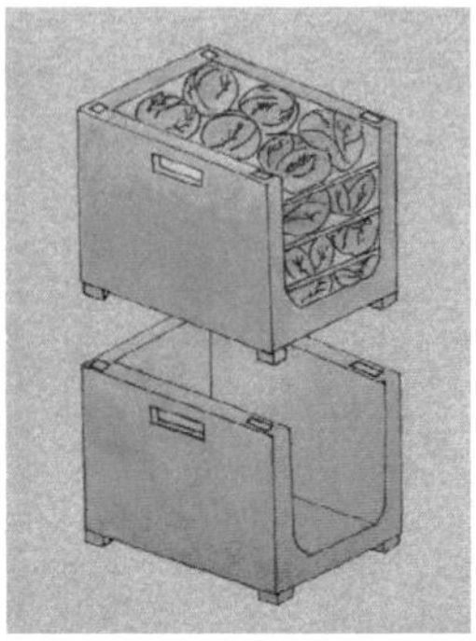

Figure 4. Low-fidelity Prototype of Cabbage Container

In traditional markets and supermarkets, cabbage diameter dimensions ranged from 14cm to 20cm, and cabbage thickness dimensions ranged from 9cm to 12cm, according to measurements. Then, a high-fidelity prototype was created from the cabbage container, taking into account the dimensions of cabbage, vehicles, and the typical transportation capacity. Figure 5 depicts a prototype of high quality for the cabbage container. A layer of EPS (expanded polystyrene) is added to containers to reduce vibrations during the distribution process (see Figure 6). In addition, the visualization of cabbage in containers was performed on the tailgate of a Mitsubishi TS120SS pickup truck (see Figure 7).

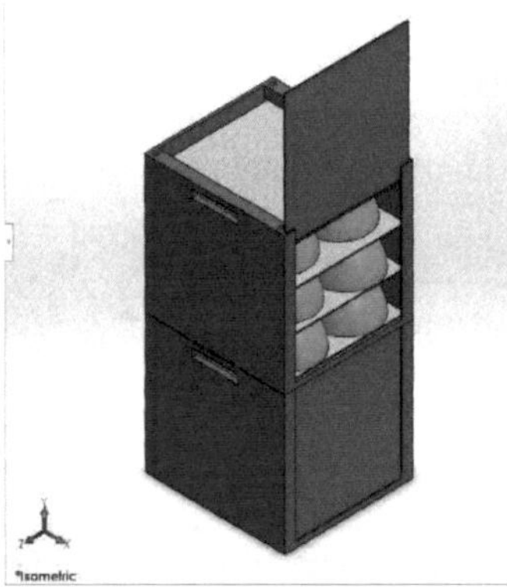

Figure 5. High-fidelity Prototype Container Filled with Cabbages

Figure 6. High-fidelity Prototype EPS for Cabbage Container

Figure 7. High-fidelity Prototype of Container Filled with Cabbages in Mitsubishi TS120SS

Next, a high-fidelity prototype of cabbage-handling instructions was created. Figures 8-10 are high-fidelity prototypes of cabbage handling instructions for farmers, intermediaries, and traders, respectively.

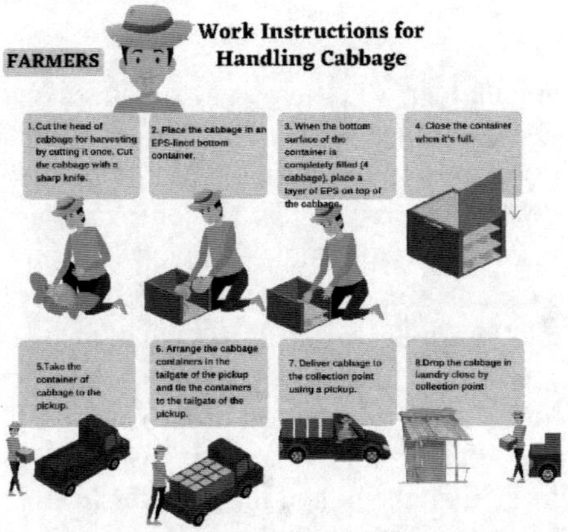

Figure 8. High-fidelity Prototype of Work Instruction for Handling Cabbage (Farmers)

Figure 9. High-fidelity Prototype of Work Instruction for Handling Cabbage (Intermediary's Employees)

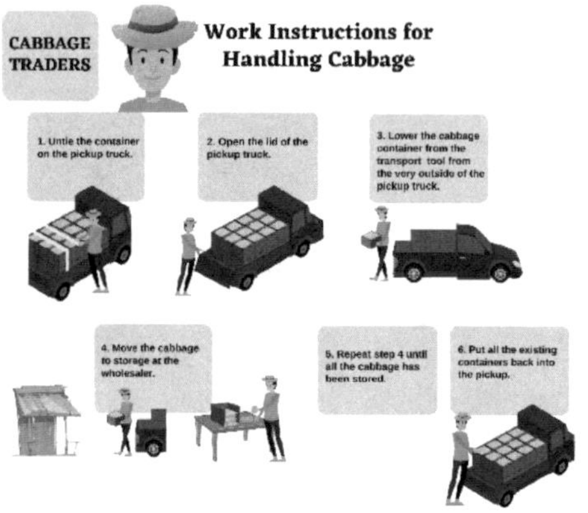

Figure 10. High-fidelity Prototype of Work Instruction for Handling Cabbage (Cabbage Traders)

The prototype of the banana container was then created using hand sketches. The image depicts a crude prototype of the banana container (see Figure 11). Observations and interviews were conducted in the market to collect data on the types of vehicles commonly used for banana distribution. Based on observations, it was determined that Mitsubishi trucks are commonly used for the distribution of bananas [9]. Additionally, the diameters of banana plantains ranged between 24 and 29 centimetres. Then, interviews with middlemen and banana merchants revealed that super bananas typically contain six to ten circles/bunches. Thus, a prototype with eight circles per bunch was created. Figure 12 depicts a prototype with high fidelity of a bunch of bananas. A layer of EPS (expanded polystyrene) is added to containers to reduce vibrations during the distribution process. The EPS layer on the banana container is illustrated in Figure 13. Figure 14 depicts a high-fidelity prototype of banana containers. In addition, container visualization was performed on the bed of the Mitsubishi Colt Diesel FE 74 HD and FE 71 PS 110 PS (see Figure 15-16). Figure 17 to Figure 19 depict the development of a high-fidelity prototype based on work instructions for farmers, intermediaries, and traders regarding the handling of bananas.

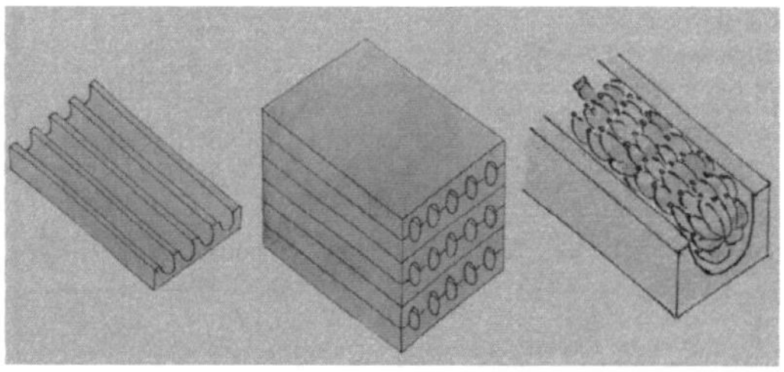

Figure 11. Low-fidelity Prototype of Banana Container

Figure 12. High-fidelity Prototype of a Banana Bunch

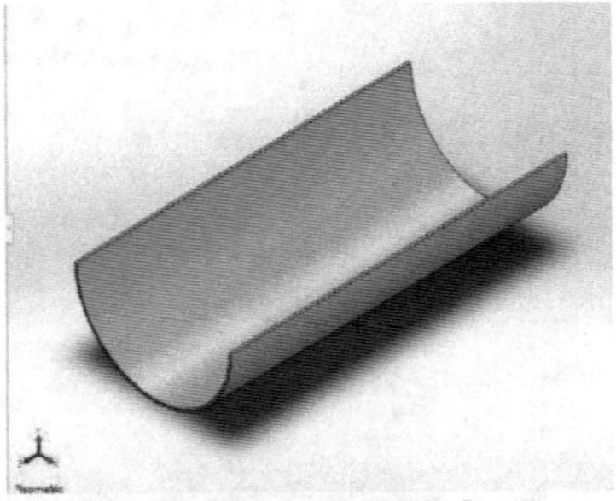

Figure 13. High-fidelity Prototype EPS for Banana Container

Figure 14. High-fidelity Prototype of Container Filled with Bananas

Figure 15. High-fidelity Prototype of Containers Filled with Bananas on Mitsubishi Colt Diesel FE 74 HD

IOP Conf. Series: Earth and Environmental Science 1256 (2023) 012007 doi:10.1088/1755-1315/1256/1/012007

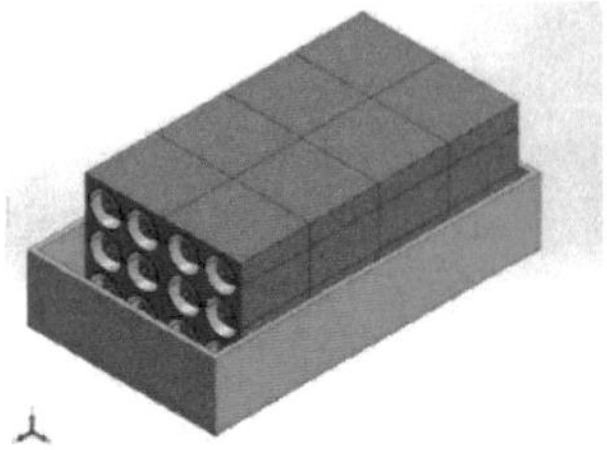

Figure 16. High-fidelity Prototype of Containers Filled with Bananas on Mitsubishi Colt Diesel FE 71 PS 110 PS

Figure 17. High-fidelity Prototype of Work Instruction for Handling Banana (Farmers)

Figure 18. High-fidelity Prototype of Work Instruction for Handling Banana (Intermediary's Employees)

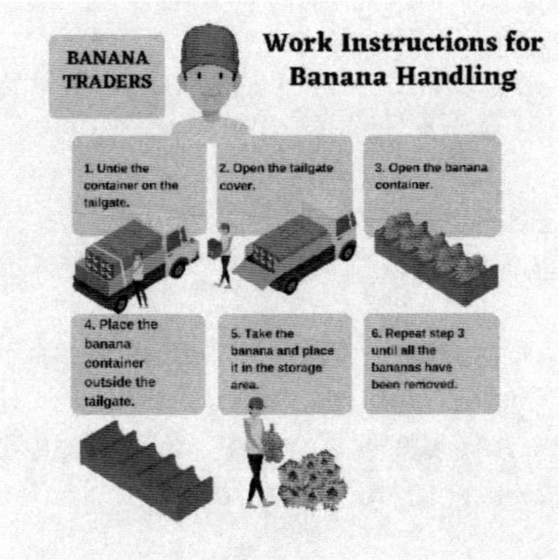

Figure 19. High-fidelity Prototype of Work Instruction for Handling Banana (Banana Traders)

Following the completion of the prototype, testing of the prototypes is conducted. The participants in the prototype testing process included intermediaries, carriers, traders, and brainwriting participants. However, it was not conducted for farmers due to government policies that restrict the movement of people due to the Covid-19 pandemic. According to the test results, all prototypes were well-received by all tested parties and were anticipated to effectively reduce food loss in horticultural products.

5. Conclusions

On the basis of the study's findings, the following conclusions can be drawn. The majority of food loss in horticultural products is due to mechanical damage resulting from the distribution process in the supply chain. To reduce food loss between farmers and traders in horticultural commodities, preventative measures, such as the use of containers during the distribution of horticultural commodities and work instructions for handling horticultural commodities, can be implemented. According to the results of testing the prototype container solution and work instructions, the use of both is anticipated to reduce food loss in horticultural commodities.

Acknowledgements:

This research was co-funded by the European Commission through the Erasmus+ CBHE project "Resolving A Societal Challenge: Interdisciplinary Approach Towards Fostering Collaborative Innovation in Food Waste Management" (IN2FOOD), project number 618717-EPP-1-2020-1-ID-EPPKA2-CBHE-JP. The European Commission's support for the production of this publication does not constitute an endorsement of the contents, which reflect the views only of the authors, and the Commission cannot be held responsible for any use which may be made of the information contained therein.

References

[1] Direktorat Jenderal Hortikultura. (2014). Rencana Strategis Direktorat Jenderal Hortikultura Tahun 2015-2019. Downloaded on April 20, 2021: https://hortikultura.pertanian.go.id/?page_id=5913#1608412626112-%2058ae8b4a-8fa7
[2] Mahayana, D. (2021). Rancangan Bahan Pengisi Kemasan Transportasi Tomat Beef. Bogor: Institut Pertanian Bogor.
[3] Chopra, S., & Meindl, P. (2016). *Supply Chain Management: Strategy, Planning, and Operation* (6th ed.). New Jersey: Pearson Education Inc.
[4] Matani, A.G., Tripathi, M.S., Doifode, S.K., and Gowardhan, S.D. (2015). Green Supply chain Management in Food Industries. *International Journal of Engineering and Technical Research (IJETR)*, 3, 261-263.
[5] Kenneth, W.G.J., Pamela, J.Z., Jeramy, M., Vikram, S.B. (2012). *Green Supply Chain Management Practices: Impact on Performance*. Bingley: Emerald Group Publishing Limited.
[6] Sarkis, J. & Dou, Yijie. (2018). *Green Supply Chain Management: A Concise Introduction.* New York.
[7] Närvänen, E., Mesiranta, N., Mattila, M., & Heikkinen, A. (2019). Food Waste Management: Solving the Wicked Problem. London: Palgrave Macmillan.
[8] Lewrick, M., Link, P., & Leifer, L. (2020). *The Design Thinking Toolbox: A Guide to Mastering the Most Popular and Valuable Innovation Methods* (1st ed.). New Jersey: Wiley.
[9] Mitsubishi. (2021). Mitsubishi T120SS. Downloaded on 4 August 2021: https://sbmgrp.co.id/mitsubishi-motors/our-cars/t120ss.

AUTHOR INDEX